GLOBAL SECURITY ENGAGEMENT
A New Model for Cooperative Threat Reduction

Committee on Strengthening and Expanding the Department of Defense Cooperative Threat Reduction Program

Committee on International Security and Arms Control

Policy and Global Affairs

NATIONAL ACADEMY OF SCIENCES
THE NATIONAL ACADEMIES

THE NATIONAL ACADEMIES PRESS
Washington, D.C.
www.nap.edu

THE NATIONAL ACADEMIES PRESS   500 Fifth Street, N.W.   Washington, DC 20001

NOTICE: The project that is the subject of this report was approved by the Governing Board of the National Research Council, whose members are drawn from the councils of the National Academy of Sciences, the National Academy of Engineering, and the Institute of Medicine. The members of the committee responsible for the report were chosen for their special competences and with regard for appropriate balance.

This study was supported by Contract No. DTRAA01-02-D0003, DO#8 between the National Academy of Sciences and the U.S. Department of Defense. Any opinions, findings, conclusions, or recommendations expressed in this publication are those of the author(s) and do not necessarily reflect the views of the organizations or agencies that provided support for the project.

International Standard Book Number-13:   978-0-309-13106-3
International Standard Book Number-10:   0-309-13106-5
Library of Congress Control Number 2009929437

Additional copies of this report are available from the National Academies Press, 500 Fifth Street, N.W., Lockbox 285, Washington, DC 20055; (800) 624-6242 or (202) 334-3313 (in the Washington metropolitan area); Internet, http://www.nap.edu

Copyright 2009 by the National Academy of Sciences. All rights reserved.

THE NATIONAL ACADEMIES
Advisers to the Nation on Science, Engineering, and Medicine

The **National Academy of Sciences** is a private, nonprofit, self-perpetuating society of distinguished scholars engaged in scientific and engineering research, dedicated to the furtherance of science and technology and to their use for the general welfare. Upon the authority of the charter granted to it by the Congress in 1863, the Academy has a mandate that requires it to advise the federal government on scientific and technical matters. Dr. Ralph J. Cicerone is president of the National Academy of Sciences.

The **National Academy of Engineering** was established in 1964, under the charter of the National Academy of Sciences, as a parallel organization of outstanding engineers. It is autonomous in its administration and in the selection of its members, sharing with the National Academy of Sciences the responsibility for advising the federal government. The National Academy of Engineering also sponsors engineering programs aimed at meeting national needs, encourages education and research, and recognizes the superior achievements of engineers. Dr. Charles M. Vest is president of the National Academy of Engineering.

The **Institute of Medicine** was established in 1970 by the National Academy of Sciences to secure the services of eminent members of appropriate professions in the examination of policy matters pertaining to the health of the public. The Institute acts under the responsibility given to the National Academy of Sciences by its congressional charter to be an adviser to the federal government and, upon its own initiative, to identify issues of medical care, research, and education. Dr. Harvey V. Fineberg is president of the Institute of Medicine.

The **National Research Council** was organized by the National Academy of Sciences in 1916 to associate the broad community of science and technology with the Academy's purposes of furthering knowledge and advising the federal government. Functioning in accordance with general policies determined by the Academy, the Council has become the principal operating agency of both the National Academy of Sciences and the National Academy of Engineering in providing services to the government, the public, and the scientific and engineering communities. The Council is administered jointly by both Academies and the Institute of Medicine. Dr. Ralph J. Cicerone and Dr. Charles M. Vest are chair and vice chair, respectively, of the National Research Council.

www.national-academies.org

COMMITTEE ON STRENGTHENING AND EXPANDING THE DEPARTMENT OF DEFENSE COOPERATIVE THREAT REDUCTION PROGRAM

DAVID R. FRANZ *(Co-Chair)*, Midwest Research Institute
RONALD LEHMAN *(Co-Chair)*, Lawrence Livermore National Laboratory
ROBERT B. BARKER, Lawrence Livermore National Laboratory (retired)
WILLIAM F. BURNS, U.S. Army War College
ROSE E. GOTTEMOELLER, Carnegie Endowment for International Peace
JOHN HAMRE, Center for Strategic and International Studies
ROBERT JOSEPH, National Institute for Public Policy
ORDE KITTRIE, Arizona State University
JAMES LEDUC, Galveston National Laboratory
RICHARD W. MIES, private consultant
JUDITH MILLER, Manhattan Institute
GEORGE W. PARSHALL, Du Pont (retired)
THOMAS R. PICKERING, Hills & Company, International Consultants
KIM SAVIT, University of Denver and private consultant

National Research Council Staff

ANNE HARRINGTON, Study Director, Committee on International Security and Arms Control
RITA S. GUENTHER, Senior Program Associate, Committee on International Security and Arms Control
BENJAMIN J. RUSEK, Senior Program Associate, Committee on International Security and Arms Control
LA'FAYE LEWIS-OLIVER, Administrative Coordinator
YOUSAF BUTT, Christine Mirzayan Science and Technology Policy Graduate Fellow
CLARK CULLY, Christine Mirzayan Science and Technology Policy Graduate Fellow
JESSICA MEISNER, Christine Mirzayan Science and Technology Policy Graduate Fellow

Preface

The success of the Department of Defense Cooperative Threat Reduction (DOD CTR) program at the end of the Cold War was not a foregone conclusion. The program to reduce weapons of mass destruction (WMD) threats was a bold idea in a time of transition and uncertainty. The risks seemed every bit as evident as the benefits. Generating action throughout an overburdened U.S. government at a time of budget cuts and change required an agility seldom found except in times of great urgency. Placing the initial responsibility for CTR in DOD and drawing upon the organizational energy of the one department most practiced at rapid mobilization of resources was a primary reason for the early success of CTR. Because DOD had, through its regional political-military responsibilities and arms control coordination, diverse skills, experienced people, and a habit of interagency networking when confronted with new challenges, the program took off. Quickly, other departments and organizations were participating as well.

While not on the scale of the Marshall Plan, history will record that the DOD CTR—or Nunn-Lugar Program—also generated great hope and stability in a time of political and economic crisis and then provided the resources for cooperation to former Cold War adversaries to enhance the well-being of all. Over time, many of its revolutionary activities became routine, and as such came to reflect all the advantages and disadvantages of being taken for granted. Bureaucratization, micromanagement, and the Washington turf wars invited rigorous measures of merit even as bigger questions were asked about the appropriateness of the program for today's circumstances. Still, scholars and policy makers continue to speculate on how bad the outcomes might have been had a CTR program not been created in 1992.

In the years ahead, we face new challenges for which tools originally developed by the DOD CTR program and then in the Departments of State and Energy, and elsewhere, may again be mobilized along with new tools that are desperately needed. Whether the long-run trend for most of the world is toward greater security, prosperity, and freedom is unclear, and many parts of the world seem destined toward turmoil and violence with a global impact. The advance and spread of dual-use technology will increasingly make access to highly destructive or disruptive technology easier and cheaper for small countries and smaller groups of nonstate actors. No "silver bullet" is likely. It is in this context that the committee believes a fresh look at DOD's CTR program is most warranted.

In its own work, the committee recognized that many CTR tools had already been modified to meet evolving circumstances. In considering how these CTR tools might be exploited further, members of the committee began to refer to proposed enhancements as CTR 2.0. This shorthand, drawn from the software industry, reflected both step-by-step problem solving and the ongoing applicability of many existing CTR approaches to new challenges and new regions. While acknowledging existing momentum, however, the term CTR 2.0 came to reflect also the committee's conclusion that a more aggressive upgrade to CTR was needed. To meet the magnitude of new security challenges, particularly at the nexus of WMD and terrorism, more and more deeply embedded cooperation involving security and threat reduction is vital. This requires more than small fixes.

Our conclusion is that a bold vision is again required and that DOD and the entire U.S. government should reexamine what CTR has already accomplished and refocus efforts to promote global security engagement in the 21st century.

Ronald F. Lehman
Co-chair

David R. Franz
Co-chair

A Note on Terminology

The committee responsible for this report discovered early in its discussions that the terminology used to describe Cooperative Threat Reduction (CTR) program activities was varied and often confusing. Some call CTR the Nunn-Lugar Program, others associate CTR primarily with the Department of Defense, and in recent years CTR has been used generically to refer to the broad group of CTR programs spread across U.S. government departments and agencies. To facilitate its discussions, the committee established a set of terms that it uses throughout this report. In considering how best to express its vision of a future version of CTR, the committee concluded that an expression borrowed from the software industry that refers to a new version of an existing program is a useful way to describe the more advanced and comprehensive approach to cooperative threat reduction that is advocated in this report.

CTR – generic reference to cooperative threat reduction

CTR 1.0 – the original cooperative threat reduction program developed at the end of the Cold War and implemented by multiple U.S. government programs in the former Soviet Union

CTR 2.0 – a set of programs and projects to be undertaken by the U.S. government, as part of a cooperative network that includes a wide range of countries, international organizations, and nongovernment partners, to prevent, reduce, mitigate, or eliminate common threats to U.S. national security and global stability that have emerged since the end of the Cold War

DOD CTR – programs under the policy direction of the secretary of defense and as defined by the annual National Defense Authorization Act. These programs are implemented by the Defense Threat Reduction Program (DTRA) and by contractors supported by DTRA

USG CTR – the set of programs across the U.S. government that are now associated with cooperative threat reduction activities

Acknowledgments

In carrying out this study, committee members and staff benefited greatly from the insights and observations of many experts and scientific colleagues. The views obtained during these discussions provided the essential input for the study. The committee expresses its gratitude for the time that these many colleagues devoted to ensuring that this study is as comprehensive and accurate as possible.

The study has also benefited from the insights of individuals who have worked with the U.S. government Cooperative Threat Reduction (CTR) programs for many years. In particular, the committee would like to thank the following individuals for their contributions during the open sessions of the committee meetings: *Joseph Benkert,* Department of Defense (DOD); *Joseph DeThomas,* Civilian Research and Development Foundation; *Joseph P. Harahan*, Defense Threat Reduction Agency (DTRA); *Mary Alice Hayward,* Department of State; *Susan Koch,* Department of State (retired); *Kenneth Luongo,* Partnership for Global Security; *Charles Lutes,* National Security Council; *Neile Miller,* Office of Management and Budget; *Mary Beth Dunham Nikitin,* Congressional Research Service; *Sharon Squassoni,* Carnegie Endowment for International Peace; *Amy Smithson,* Monterey Institute of International Studies; *William Steiger,* Department of Health and Human Services; *James Tegnelia,* DTRA (retired); *Charles Thornton*, University of Maryland; *William Tobey*, Department of Energy (retired); and *Elizabeth Turpen*, the Henry L. Stimson Center.

The committee was also fortunate to have the opportunity to interact with individuals who worked directly on important national security efforts in Albania and Libya, whose input provided key insights. The committee is grateful to *Karin Look*, Department of State; *Donald Mahley*, Department of State (retired); *Steven Saboe*, Department of State; *Kenneth Ward*, Department of State; and *Kenneth Myers*, Senate Foreign Relations Committee, for the time they spent speaking with various committee members.

As DOD begins to look at expanding its CTR activities globally, the committee thought that it was important to understand the relationship between what DOD CTR does programmatically and the missions of the Unified Combatant Commands. The committee discussed sets of questions with the senior leadership and staff of three commands and is grateful to Major General *Vern T. Miyagi* at U.S. Pacific Command; Major General *Paul G. Schafer* at U.S. European Command; and Vice Admiral *Robert Moeller* and Ambassador *Mary Yates* at U.S. Africa Command for the time and attention they provided. The committee would also like to thank Lieutenant Colonel *Mark Drabecki*, Lieutenant Colonel *Charles Tennyson*, and Lieutenant Colonel *Shannon McCoy* for arranging the consultations.

This study has been reviewed in draft form by individuals chosen for their diverse perspectives and technical expertise in accordance with procedures approved by the National Research Council's Report Review Committee. The purpose of these independent reviews is to provide candid and critical comments that assist the institutions in making this study as sound as possible and ensure that the study meets institutional standards for objectivity and accuracy. The review comments and draft manuscript remain confidential to protect the integrity of the process. We wish to thank the following individuals for their review of this report: *R. Stephen Berry*, University of Chicago; *Ambassador Linton Brooks*, Department of Energy (retired); *Mona Dreicer*, Lawrence Livermore National Laboratory; *Margaret Hamburg*, Nuclear Threat Initiative; *Susan Koch*, Department of State (retired); *Kenneth Luongo*, Partnership for Global Security; *Adel A. F. Mahmoud*, Princeton University; *Sharon Squassoni*, the Carnegie Endowment for International Peace; and *Paul Walker*, Global Green.

Although the reviewers listed above have provided many constructive comments and suggestions, they were not asked to endorse the publication, nor did they see the final draft of the report before its release. The review of the report was overseen by *Stephen E. Fienberg*, Carnegie Mellon University, and *Alvin Trivelpiece*, Oak Ridge National Laboratory (retired). Appointed by the National Academies, they were responsible for making certain that an independent examination of this publication was carried out in accordance with institutional procedures and that all review comments were carefully considered. Responsibility for the final content of this report rests entirely with the authoring committees and institutions.

Contents

Executive Summary		1
Overview		5
Introduction		17
1	The Evolution of Cooperative Threat Reduction (CTR)	21
2	Cooperative Threat Reduction in the 21st Century: Objectives, Opportunities, and Lessons	39
3	The Form and Function of CTR 2.0: Engaging Partners to Enhance Global Security	69
4	The Role of the Department of Defense in CTR 2.0	99
5	CTR 2.0: Implementation Checklist	117
List of Acronyms		123

Appendixes

A	H.R. 1585: National Defense Authorization Act for Fiscal Year 2008	129
B	Biographical Sketches of Committee Members	133

C	Department of Defense Cooperative Threat Reduction Program History: References	139
D	List of Committee Meetings and Speakers	141
E	The Evolution of U.S. Government Threat Reduction Programs	143
F	Nunn-Lugar Scorecard	149
G	The G8 Global Partnership: Guidelines for New or Expanded Cooperation Projects	151
H	A Comparison of the Characteristics of Six Weapons Systems from the Perspective of a State or Terrorist Organization	155
I	Department of Defense Cooperative Threat Reduction Programs	159
J	Congressional Guidelines and Corresponding Findings and Recommendations	163

Executive Summary

In the National Defense Authorization Act of 2008, Congress directed the National Academy of Sciences to recommend ways to strengthen and expand the Department of Defense Cooperative Threat Reduction (DOD CTR) program, including the development of new initiatives. In early consultations with congressional staff, the committee appointed to author this report was also asked to come to its own judgment about the future of the DOD CTR program. **The committee concludes that expanding the nation's cooperative threat reduction programs beyond the former Soviet Union, as proposed by Congress, would enhance U.S. national security and global stability. In this report the committee proposes how this goal can best be achieved.**

The committee recommends that the **DOD CTR program should be expanded geographically, updated in form and function according to the concept proposed in this report, and supported as an active tool of foreign policy by engaged leadership from the White House and the relevant cabinet secretaries (Recommendation 1-1).**

As requested by Congress, this report identifies a number of promising program areas that can be the basis for expanded activities across a number of regions and countries. However, future efforts to enhance global security must be part of a broader, integrated set of programs. To meet the magnitude of new security challenges, particularly at the nexus of weapons of mass destruction and terrorism, a new model is needed that will draw on a broader range of partners and require more flexibility than current programs have. **The White House, working across the executive branch and with Congress, should engage a broader range of partners in a variety of roles to enable a new program model to enhance global security. At a minimum this will require:**

- *Becoming more agile, flexible, and responsive*
- Cultivating *additional domestic and global partners* to help meet its goals
- Building mutually beneficial *relationships* that foster *sustained cooperation* (Recommendation 2-1)

Strong White House leadership will be necessary to achieve the integration needed for a new program model, but no new effort will succeed without the active and committed support of cabinet secretaries and other senior officials from all relevant agencies. **The new global security engagement effort should be directed by the White House through a senior official at the National Security Council and be implemented by the Departments of Defense, State, Energy, Health and Human Services, Agriculture, and other relevant cabinet secretaries (Recommendation 3-1).**

This new CTR program strategy will need to take into account resources available across the government and through nongovernment and international partners. **Domestically, the program should include a broad group of participants, including government, academe, industry, nongovernmental organizations and individuals, and an expanded set of tools, developed and shared across the U.S. government (Recommendation 3-1a). Internationally, the program should include multilateral partnerships that address both country- and region-specific security challenges, as well as provide support to the implementation of international treaties and other security instruments aimed at reducing threat, such as the G8 Global Partnership, the Proliferation Security Initiative, United Nations Security Council Resolution 1540, and the Global Initiative to Combat Nuclear Terrorism (Recommendation 3-1b).**

A new version of CTR—a CTR 2.0—will face very different security challenges than those that inspired the original program nearly 20 years ago. Forging broad new partnerships to implement sustainable programs that employ hard and soft capabilities and are tailored to specific countries or regions will energize and strengthen global security efforts and result in tangible and intangible benefits to national security. It is essential to develop meaningful program metrics that highlight program impact, acknowledge the value to national security of intangible program results, incorporate partner metrics into the overall evaluation of programs, and link metrics to program selection criteria. **The Executive Branch and Congress need to recognize that personal relationships and professional networks that are developed through USG CTR programs contribute directly to our national security and that new metrics should be developed to reflect this (Recommendation 3-2).**

U.S. government bureaucracy is difficult for international partners to understand and often delays project implementation for many months, appearing to our partners as reluctance to work with them. Several specific measures can make the next generation of global security engagements more efficient, timely, and valuable. This will lead to greater confidence, transparency, and, ultimately, enhanced national security. **The legislative framework, funding mechanisms, and program leveraging opportunities should be structured to support more effective threat reduction initiatives across DOD, other USG departments and agencies, international partners, and NGOs (Recommendation 3-3).**

a) Program planning should be developed out of a strategic process and be matched by a *strategic budget process* that produces a multiyear budget plan and distributes funding across agencies based on agency ability to respond to program requirements. As needed, agency legislative authorities should be revised to include a national security dimension (Recommendation 3-3a).

b) Congress should **provide** *comingling authority* to all agencies implementing programs under CTR 2.0 as a way to encourage other partners to contribute funds to global security engagement efforts (Recommendation 3-3b).

c) To maximize the effectiveness of CTR 2.0, the DOD CTR legal frameworks and authorities should be reassessed. DOD should undertake a **systematic study of the** *CTR Umbrella Agreement* **protection provisions**, what purposes they serve in which circumstances, whether there might be less intrusive means of accomplishing the provisions' goals, and when the provisions are necessary in their present form. In addition, all USG CTR programs should identify legal and policy tools that can promote the sustainability of U.S.-funded CTR work and provide greater implementation flexibility (Recommendation 3-3c).

d) Congress should grant DOD **limited** *"notwithstanding" authority* for the CTR program—perhaps a maximum of 10 percent of the overall annual appropriation and subject to congressional notification—to give the program the additional flexibility it will need in future engagements (Recommendation 3-3d).

The CTR concept began almost two decades ago and programs should periodically be reviewed and evaluated. **The Secretary of Defense should direct the review and reformulation of the DOD CTR program in support of the new model of global security engagement proposed in this report and work with the White House, Secretary of State, Secretary of Energy, and other cabinet and agency officers to ensure full coordination and effective implementation of DOD programs in this new model. The review should also include broader military components, including the Unified Combatant Commands, the full set of programs in DTRA, DOD health and research programs, and other DOD assets (Recommendation 4-2).**

Overview

The world has changed. Lines are now blurred: lines between nations, regions, and peoples; lines between disciplines, tools, and applications of chemistry, physics, and biology; and lines between the use of technologies for good or evil. As capabilities have spread around the globe, small groups and individuals have gained access to instruments of harm that once belonged exclusively to nation states. When vast armies threatened, the United States found tools to reduce the threat. In recent years, the United States has had to shift the emphasis of its hard-tool set from heavy artillery and armor to more agile and flexible light infantry, special operators, and precise delivery of kinetic weapons. It must now do the same with its soft tools and apply them with similar agility and precision. This transformation will require enlightened and engaged leadership; effective communication across the U.S. government; a networked culture of cooperation among like-minded nations; and the engagement of new partners in academe and industry and with nongovernment organizations (NGOs).

The National Defense Authorization Act of 2008 called for a National Academy of Sciences study that would assess new initiatives for the Department of Defense Cooperative Threat Reduction (DOD CTR) program, particularly in the Middle East, Asia, and the Democratic People's Republic of Korea, and identify options and recommendations for strengthening and expanding the CTR program.[1] Senators Sam Nunn and Richard Lugar crafted the original CTR program as an innovative response to threats posed by the collapse of the Soviet Union; similar creativity is needed now to develop an enhanced program that involves new players, new places, and new programs.

When the Soviet Union fell, a disheartened and dispersed military force remained in place, still responsible for tens of thousands of nuclear weapons,

---

[1] See Appendix A for the full text of the legislation.

hundreds of tons of chemical weapons; and a massive biological weapons research, development, and production infrastructure. Much of the remaining weapons of mass destruction (WMD) capability existed in closed cities and limited access areas, many of which were known only by postal codes and never appeared on official Soviet maps. The potential loss of weapons and the vulnerability of weapons materials and expertise drove a sense of urgency.

U.S. negotiators arrived in Moscow with no specific plan and through constructive discussions between senior military officers, officials, and technical experts, the Cooperative Threat Reduction program was born. The initial focus was to assist the Newly Independent States (NIS) of the former Soviet Union (FSU), particularly those in which nuclear weapons were located.

The DOD CTR program was initially authorized by Public Law 102-228. The law defined three primary program objectives: (1) assist the former Soviet states to destroy nuclear, chemical, and other weapons; (2) transport, store, disable, and safeguard weapons in connection with their destruction; and (3) establish verifiable safeguards against the proliferation of such weapons. In 1992, these objectives were expanded to include dismantling missiles and missile launchers; destroying destabilizing conventional weapons; preventing diversion of weapons-related scientific expertise; establishing science and technology centers; facilitating demilitarization of defense industries and converting military capabilities and technologies; and expanding military-to-military and defense contacts.

The DOD CTR program had few precedents to guide its initial development, but there was a sense of urgency that was shared by leaders in both Russia and the United States, in some cases for different reasons. Russia's new leaders were interested in remaining the sole nuclear power in the region, but also recognized that foreign financial assistance would be critical to consolidate, safeguard, and in some cases dismantle weapons systems as well as to help the country through a turbulent economic period. U.S. leaders were concerned about the potential threat from four new nuclear states, about accountability for any U.S. assistance provided for threat reduction, and how to ensure that assistance provided was not used to sustain or enhance former Soviet weapons capabilities.

DOD policies, procedures, and rules developed to implement its CTR program were complex, and the process of putting agreements into place to govern the new program activities were unfamiliar to the leaders of the NIS. In the United States, some individuals in Congress were unconvinced that the program was in U.S. national security interests and saw the program more as foreign assistance. Despite a long record of CTR accomplishments, the challenge of demonstrating the national security benefits of CTR 2.0 will also require an ongoing set of consultations between the executive and legislative branches to ensure that members of Congress and their staffs understand the program's strategy and approaches.

During the 15 years that followed passage of the Nunn-Lugar legislation, DOD invested nearly $7 billion to safeguard and dismantle vast stockpiles of nuclear, chemical, and biological weapons, or related materials and delivery systems, within a framework of cooperative engagement.[2] From the beginning, the DOD CTR program worked closely with sister programs in the Department of State and Department of Energy, forming a set of U.S. government (USG) CTR efforts. These programs have evolved over the years, often in response to congressional directions, restrictions, prohibitions, or preferences.

Much of DOD's CTR engagement has been through large integrating contractors that have implemented expensive and extended engineering demilitarization or construction projects. As many of the engineering projects near completion and as U.S.-Russian relations evolve, the volume of program activity in Russia has contracted significantly, from a budget of nearly $375 million in 1999, to slightly more than $150 million in 2008.[3] Some DOD CTR work, especially in biological nonproliferation, is expanding beyond Russian states in the FSU under the Biological Threat Reduction Program; constructing effective border security and export control systems also continues throughout the region. But the emphasis has shifted from destroying and securing weapons facilities and engaging former weaponeers to increasing security through building detection and disease surveillance capability, whether for detecting biological events or stopping traffickers. Likewise, the metrics of success for USG CTR programs have been changing from "weapons and systems destroyed" to "nonproliferation capabilities enhanced." These metrics need to evolve further to reflect the importance of intangible as well as tangible program outcomes, and to better reflect program impact in partner countries.

Intense oversight by Congress and more than 40 Government Accountability Office reports on the DOD CTR program activities were driven by an early sense of caution regarding the potential that these programs might contribute to helping Russia enhance its military power. These controls may have provided management security, but they also resulted in a bureaucratic burden that, according to one person closely involved, "almost monitored the program to death." Officials in partner countries as well as in U.S. agencies were frustrated by implementation delays that often were interpreted as U.S. reluctance to cooperate. A new approach that highlights program transparency is needed to provide the assurance that public funds are being spent responsibly, while allowing for program flexibility.

Since its inception, the DOD CTR program has made significant contributions to reducing the spread of nuclear, chemical, and biological weapons. The

---

[2] Amy Woolf. 2008. *Nonproliferation and Threat Reduction Assistance: U.S. Programs in the Former Soviet Union*. Washington, D.C.: Congressional Research Service. 11 pp. Accessed at http://fas.org/sgp/crs/nuke/RL31957.pdf, May 19, 2009.
[3] Ibid.

program is justifiably proud of the tangible results it has achieved—deactivating thousands of warheads, destroying intercontinental ballistic missiles and their silos, dismantling strategic submarines and bombers, neutralizing chemical weapons, and destroying or converting biological weapons production facilities; redirecting former weapons scientists and engineers; and initiating biological surveillance efforts in Russia and the NIS. These activities have also had intangible results in the hundreds or thousands of personal relationships among scientists, engineers, military officers, and government officials in the FSU and the United States. These relationships support frank and open communication despite periods of bilateral tensions.

The National Research Council committee that authored this report concludes that U.S. national security and global stability would be enhanced by expanding the nation's cooperative threat reduction programs beyond the former Soviet Union and readdressing their form and function. To this end, the committee has looked broadly at how the original cooperative threat reduction programs—or CTR 1.0—can be upgraded and improved to create a new approach to global security engagement, which we call CTR 2.0 (see Box O.1).

In this study, the committee explored how the CTR concept can best be applied to contemporary WMD and terrorist threats on a global scale. Although the end of the Cold War presented a diverse and complex set of challenges, the issues were largely concrete and identifiable. But the threats of the 21st century are fundamentally different. The rapid globalization of com-

---

BOX O.1
What Is CTR 2.0?

CTR 2.0, an expression borrowed from the software industry, refers to a more advanced and comprehensive approach to cooperative threat reduction. It comprises a set of programs and projects undertaken by the United States, as part of a cooperative network that includes a wide range of countries, international organizations, and nongovernment partners, to prevent, reduce, mitigate, or eliminate common threats to U.S. national security and global stability that have emerged in particular since the end of the Cold War. The preferred mechanism and long-term goal for the cooperation is partnership, which means that the countries participating should be ready to share responsibilities for project definition, organization, management, and financing according to a rational division of labor, capacity (including budget capacity), or technical capability. Although CTR 2.0 engagements may have to begin under less than ideal circumstances, the goal for countries engaged under CTR 2.0 is shared responsibility through engagement and partnership. CTR 2.0 should be capable of rapid response as well as longer-term programmatic engagement.

munications, transportation, and knowledge allow threats to be networked, agile, adaptable, and difficult to quantify. New tools and programs are needed to respond to these threats. In the committee's view, a fundamentally different approach to CTR is required.

The risks that the United States faces today are no longer reduced significantly by friendly neighbors to the north and south and vast oceans to the east and west. The world is smaller than it was in 1992. Ignoring globalization is not an option, whether in economics, public health, combating terrorism, or reducing the threat of WMD. While our technological and military capabilities will continue to play an essential role, engagement is also one of the most important tools in the national security arsenal. Forging partnerships will require strong and creative leadership from the White House; dedicated and attentive leadership in government departments and agencies; and updated, integrated, and effectively coordinated CTR programs. Relevant, sustainable CTR 2.0 programs that employ hard and soft capabilities and are tailored to a specific country or region will energize and strengthen CTR 2.0 and result in tangible and intangible national security benefits.

This report does not look comprehensively at all opportunities that might be available for the application of DOD CTR as an element of CTR 2.0, but during the committee's deliberations and in its discussions with experts, several program needs and opportunities were identified. These include some activities already associated with DOD CTR, such as promoting biological safety, security, and surveillance programs; supporting the implementation of the Chemical Weapons Convention; and enhancing border security assistance that can be applied to new regions and countries, such as the Middle East, Asia, and Africa. New program areas were also identified, such as promoting the implementation of the United Nations Security Council Resolution (UNSCR) 1540 and promoting chemical safety and security. Following the CTR 2.0 model, all of these activities would fall under coordinated strategic guidance and engage a broad range of partners.

The CTR 2.0 model envisioned by the committee and developed in this report can be summarized through the report's recommendations identified below. Each chapter concludes with that chapter's findings and recommendations.

RECOMMENDATIONS

For approximately $400 million per year over the past 15 years,[4] the DOD CTR program has demonstrated that direct engagement can roll back and eliminate programs to design and produce nuclear, chemical, and biological weapons. For less than a total of $7 billion over 15 years, these programs have deactivated thousands of nuclear warheads, supported chemical weapons destruction,

---

[4] Ibid.

transformed former biological weapons facilities, redirected former weapons scientists, and fostered communication among former enemies. In addition to WMD dismantlement, destruction, consolidation, and security, these programs have also increased transparency and helped foster higher standards of conduct and operations and the development of a security culture, as well as collaboration between civilian and military experts of the United States and the former Soviet states. These and other engagement activities have directly and indirectly enhanced U.S. national security and global security and stability.

The global spread of advanced technologies, the rise of asymmetric warfare, and the growing global interdependence of peoples, economies, and politics have made discerning an adversary's intentions more important than ever before. The footprints of weapons-producing laboratories and the size of today's "strategic" weapons grow smaller every day and their "delivery systems" may be individuals or commercial cargo carriers. Hence, discovering weapons activities is far more challenging now than it was when CTR began. Having the capacity to evaluate intentions will be key and depends on communicating directly with people in places where such capabilities exist. If the U.S. government engages only where it knows weapons are being produced, it will engage neither as much as it should, nor where it must.

CTR 1.0 relied heavily on DOD for its implementation. But responding to 21st-century threats demands a much broader range of capabilities, expertise, and "faces." In some instances, a military face may not always be most effective, as suggested by the difficulties DOD had in its efforts to engage Russia in countering biological threats. In addition, CTR 2.0 will support the implementation of bilateral and international nonproliferation, arms control, and counterterrorism agreements, and innovative initiatives and activities such as the Proliferation Security Initiative, the Global Initiative to Combat Nuclear Terrorism, and UNSCR 1540. To succeed, CTR 2.0 will require sustained White House leadership and the full cooperation of cabinet secretaries and agency heads.

Recommendation 1-1: The DOD CTR program should be expanded geographically, updated in form and function according to the concept proposed in this report, and supported as an active tool of foreign policy by engaged leadership from the White House and the relevant cabinet secretaries.

CTR 1.0 was designed to deal with yesterday's strategic weapons. The DOD CTR program has evolved into a complex enterprise in which what is "best" for a foreign partner may be decided without that partner's input. Many program efforts depend on the U.S. contracting process that can take years to complete, and initiating even small projects can take many months. In the new, more nuanced security environment, the traditional programs and their metrics will need to be complemented by new, more flexible efforts and measures of success.

At the heart of CTR 2.0 is a presumption of cooperation. Programs must have roots in the partner country and partners should be involved in a program's design, planning, and implementation. Targeted engagements supported by the Department of Health and Human Services or the Department of Agriculture or the Environmental Protection Agency may complicate terrorist efforts to exploit the resources, capabilities, or sympathies of a population more effectively than the multimillion dollar construction projects that characterized CTR 1.0.

Recommendation 2-1: The White House, working across the executive branch and with Congress, should engage a broader range of partners in a variety of roles to enable CTR 2.0 to enhance global security. At a minimum this will require

- *Becoming more agile, flexible, and responsive*
- **Cultivating** *additional domestic and global partners* to help meet its goals
- **Building mutually beneficial** *relationships* that foster *sustained cooperation*

CTR 1.0 engagements have become a portfolio of loosely coordinated actions implemented by departments and agencies across the USG. For CTR 2.0 to be effective, its form must match its functions. Strong White House leadership and sustained engagement at senior levels of all departments and agencies that contribute will need to become the norm. The National Security Council (NSC) and the Homeland Security Council (HSC) are already collaborating in biological global security engagement in an effort known as the "United States Bioengagement Strategy." This mechanism brings together representatives from the entire program spectrum, regardless of whether agencies have a legislatively mandated national security mission, initially to exchange program information and subsequently to fashion government-wide engagement strategies for several countries. The purpose is to "promote coordination," find "gaps in current activities," and help stakeholders understand "which programs should be developed or expanded." A valuable outcome of this effort will be sharing information systematically among agencies about ongoing activities in a specific country or region. This effort may be a useful model for coordination in other areas.

Once interagency coordination becomes routine, broader collaborations should be sought with a range of domestic and international partners. This will allow the U.S. government to match policy objectives with the most effective tools across such activities as WMD dismantlement and engagement of weapons specialists, export control and border security, regulatory assistance and reform, and security partnerships. The long-term goal should be to build networks of expertise capable of addressing threats and moving from assistance to partnership.

Recommendation 3-1: CTR 2.0 should be directed by the White House through a senior official at the National Security Council and be implemented by the Departments of Defense, State, Energy, Health and Human Services, and Agriculture, and other relevant cabinet secretaries.

Recommendation 3-1a: Domestically, CTR 2.0 should include a broad group of participants, including government, academe, industry, nongovernmental organizations and individuals, and an expanded set of tools, developed and shared across the U.S. government.

Recommendation 3-1b: Internationally, CTR 2.0 should include multilateral partnerships that address both country- and region-specific security challenges, as well as provide support to the implementation of international treaties and other security instruments aimed at reducing threat, such as the G8 Global Partnership, the Proliferation Security Initiative, United Nations Security Council Resolution 1540, and the Global Initiative to Combat Nuclear Terrorism.

Professional colleagues—friend or foe—throughout the world respect intellect and technical competence. Relationships provide opportunities for communication, access, and even transparency in times of great national tension, and may be one of the most important achievements of CTR programs. From the early DOD CTR senior-level military exchanges to recent collaborations in disease surveillance, close relationships formed around professional interactions persist, even where tensions between countries are heightened. Because of the fundamental change in the nature of threats and the pace at which events occur, the ability to communicate directly with a specialist in another country on a regular basis to discuss an emerging disease with a fellow public health official or a terrorist attack in his or her country has greater national security significance today than it did when CTR was founded. CTR 2.0 should value and foster such ties and find appropriate metrics to reflect their value to national security.

Recommendation 3-2: The executive branch and Congress need to recognize that personal relationships and professional networks that are developed through USG CTR programs contribute directly to our national security and that new metrics should be developed to reflect this.

Congress has done much over the years to amend legislation in ways that allow USG CTR programs to operate more broadly and effectively, but some legal and policy underpinnings of the current CTR 1.0 efforts are cumbersome and dated and often diminish the value of programs. Although the DOD CTR authorizing legislation has undergone some fundamental, positive changes, several issues need to be addressed if CTR 2.0 is to operate optimally. Some of these may require congressional action; others may be resolved by executive branch action.

The committee believes that these changes will require regular consultation between the legislative and executive branches. Senators Nunn and Lugar have been strong and vocal champions of CTR 1.0, and without their vision and commitment the program would not exist. But CTR 2.0 is an even more complex and possibly larger endeavor; it, too, will require congressional champions and a forum in which both they and critics can discuss the many issues that will inevitably arise. The committee's observations about the need for stronger leadership, coordination, and cooperation in the executive branch apply equally to Congress.

International CTR partners have little or no understanding of the U.S. government or its processes. Bureaucratic machinations, which can delay project implementation for many months, can appear to U.S. partners as reluctance to work with them. DOD in particular must reconsider its approach to umbrella agreements, geographic limitations, and the metrics by which it measures program success. Comingling authorities are needed to make it easier to work together across countries and organizations. Contracting procedures need to be streamlined, and a project's sustainability should be considered before engagement, not as an afterthought. Giving CTR 2.0 leaders, decision makers, and implementers appropriate legal and policy authorities will make engagements more efficient, timely, and valuable, and give partners a more positive perception of our commitment. This will lead to greater confidence, transparency, and, ultimately, enhanced national security.

Recommendation 3-3: The legislative framework, funding mechanisms, and program leveraging opportunities should be structured to support more effective threat reduction initiatives across DOD, other U.S. government departments and agencies, international partners, and NGOs.

Recommendation 3-3a: Program planning should be developed out of a strategic process and be matched by a *strategic budget process* that produces a multiyear budget plan and distributes funding across agencies based on agency ability to respond to program requirements. As needed, agency legislative authorities should be revised to include a national security dimension.

Recommendation 3-3b: Congress should provide *comingling authority* to all agencies implementing programs under CTR 2.0 as a way to encourage other partners to contribute funds to global security engagement efforts.

Recommendation 3-3c: To maximize the effectiveness of CTR 2.0, the DOD CTR legal frameworks and authorities should be reassessed. DOD should undertake a systematic study of the *CTR Umbrella Agreement* protection provisions, what purposes they serve in particular circumstances, whether there might be less intrusive means of accomplishing the provisions' goals, and when the provisions are necessary in their present form. In addition, all USG CTR programs should identify legal and policy tools that can promote the sustainability of U.S.-funded CTR work and provide greater implementation flexibility.

Recommendation 3-3d: Congress should grant DOD limited *"notwithstanding" authority* for the CTR program perhaps a maximum of 10 percent of the overall annual appropriation and subject to congressional notification to give the program the additional flexibility it will need in future engagements.

CTR 1.0 began as a means of assisting partners in the FSU when there were few other options. The world is unlikely to confront a similar situation again and new challenges will vary regionally and from state to state. The committee believes strongly that CTR 2.0 must be characterized by a spirit of seamless cooperation, both with the U.S. engagement team and, when possible, with the country engaged. White House guidelines for CTR 2.0 will help evaluate the best initial engagement options and the agencies that are most appropriate to the tasks at hand. Under CTR 2.0, less-developed countries may still require financial support, but they may be able to contribute in kind and should still be engaged as partners in program planning, development, and implementation. In other cases, the partners will require technology or expertise, with little cost to the U.S. government. At times the U.S. government may be neither welcome nor able to assist, but can team up with others who do have the ability to respond. This may be particularly true when DOD—or other U.S. agencies—are unwelcome, at least initially.

Recommendation 4-1: As CTR 2.0 engagement opportunities emerge, the White House should determine the agencies and partners that are best suited to execute them, whether by virtue of expertise, implementation capacity, or funding.

DOD understands the history and culture of threat reduction engagement as traditionally defined, but needs to evaluate how to engage in the future. The secretary of defense should take the lead by initiating an in-depth review, evaluation, and reformulation of CTR 1.0 to incorporate all the relevant tools within DOD. This should be done in close collaboration with current and potential CTR 2.0 partners within the U.S. government and with full engagement of responsible leadership in the White House. The review should seek to better understand historical activities that have limited legislative, operational, or geographic restrictions. The evaluation of current programs and activities viewed through the prism of CTR 2.0, the application of lessons learned to new approaches, and the incorporation of tools and partners not previously considered will demonstrate the value of the department's capabilities to the future of global security engagement.

Because the focus of CTR 1.0 was Russia and the former Soviet Union, the Unified Combatant Commands, other than the European Command, have not been involved in programs, are not part of the planning process, and even are unaware of many CTR 1.0 activities in their areas of responsibility. If CTR 2.0 is to operate globally, the Unified Commands logically should contribute to program planning and be aware of implementation. One program that is well suited to the challenges identified to the committee by the Unified Commands is the Defense and Military Contacts Program. The program is currently funded by DOD CTR, but could much better serve global security engagement activities than it does now. Military-to-military activities and the engagements that exist in combatant commands, for example, could be coordinated with the interagency under CTR 2.0, allowing commands to be aware of programs that could support these missions. Military-to-military engagements offer opportunities to initiate specific relationships and capacity building that supports the broader goals of CTR 2.0.

Recommendation 4-2: The secretary of defense should direct the review and reformulation of the DOD CTR program in support of CTR 2.0 and work with the White House, secretary of state, secretary of energy, and other cabinet and agency officers to ensure full coordination and effective implementation of DOD programs in CTR 2.0. The review should also include broader military components, including the Unified Combatant Commands, the full set of programs in the Defense Threat Reduction Program, DOD health and research programs, and other DOD assets.

Existing CTR programs have incrementally evolved toward CTR 2.0 over the years, but a more specific transition plan is needed. As the committee proposes some major changes, it applauds the interagency effort led by the NSC and the HSC to develop a bioengagement strategy, which epitomizes the spirit of CTR 2.0. The NSC-HSC team should expand its effort by reaching out to traditional and nontraditional partners, possibly focusing on one country as a test case. Once the system has been established and the mechanisms have been defined, other working groups could develop similar models, working with different challenges in different countries and regions, to create the program we call CTR 2.0.

Recommendation 4-3: A plan for the evolution of CTR 1.0 to CTR 2.0 should take into account the congressional principles enumerated in the legislation authorizing this report, as well as existing USG CTR initiatives. The White House should review National Security Council–Homeland Security Council coordination in bioengagement as a possible model for other programs as it develops a transition plan.

BOX O.2
Statement by Senator Richard Lugar

We must take every measure possible in addressing threats posed by weapons of mass destruction. We must eliminate those conditions that restrict us or delay our ability to act. The United States has the technical expertise and the diplomatic standing to dramatically benefit international security. American leaders must ensure that we have the political will and the resources to implement programs devoted to these ends.

SOURCE: Richard Lugar. We Must Take Every Measure to Address WMD Threats, Lugar Says. Press Release. December 2, 2008. Accessed at http://lugar.senate.gov/record.cfm?id=305375& on May 4, 2009.

Introduction

The National Defense Authorization Act of 2008 (Public Law 110-181, Title XIII, Section 1306) includes a provision calling for the secretary of defense to enter into an agreement with the National Academy of Sciences (NAS) to undertake a study on strengthening and expanding the Department of Defense Cooperative Threat Reduction (DOD CTR) program. Congress stipulated that the study should include an assessment of new CTR initiatives and identify options and recommendations for strengthening and expanding the CTR program. Section 1306 identifies regions that should be considered, noting in particular the potential for DOD CTR programs in the Middle East and Asia, and for activities related to the denuclearization of the Democratic People's Republic of Korea (DPRK). The legislation also includes a statement of the sense of Congress that new programs should be implemented following several principles. As the committee proceeded with its work engaging both the Congress and the executive branch, it became clear that an effective DOD CTR program for the future could only be understood in the context of a comprehensive and synergistic government-wide CTR effort. Thus, the committee has sought to address the broader issues necessary to fulfill its mandate with respect to the DOD CTR program.

In March 2008, the National Research Council (NRC), acting on behalf of the NAS, entered into a contract with the Defense Threat Reduction Agency (DTRA), acting on behalf of DOD, to carry out this study. The resulting report sets forth the findings and recommendations of the Committee on Strengthening and Expanding the Department of Defense Cooperative Threat Reduction Program established by the NRC to undertake the study. (See Appendix B for biographical information on the committee members.)

Also included in the National Defense Authorization Act of 2008 (Section 1308) was a provision calling for an NAS study of how the DOD CTR Biological Threat Reduction Program might be applied to developing countries. In response, a separate report has been prepared by the NRC, entitled *Countering Biological Threats: The Important Role of the Department of Defense's Nonproliferation Program Beyond the Former Soviet Union*. A separate NRC committee was responsible for that report, which was released in February 2009. Although that report focuses specifically on issues in the biological field, there is some overlap with this report. The two reports are intended to be complementary, but each was produced independent of the other.

STATEMENT OF TASK

This study responds to the task set forth in the legislation and in the subsequent contract between NRC and DTRA (see Appendix A for full legislation):

1. An assessment of new CTR initiatives to include at a minimum

- Programs and projects in Asia and the Middle East; and
- Activities relating to the denuclearization of the DPRK.

2. An identification of options and recommendations for strengthening and expanding the CTR program.

New initiatives should

- Be well coordinated with the Department of Energy, the Department of State, and any other relevant U.S. government agency or department;
- Include appropriate transparency and accountability mechanisms, and legal frameworks and agreements between the United States and CTR partner countries;
- Reflect engagement with nongovernmental experts on possible new options for the CTR program;
- Include work with the Russian Federation and other countries to establish strong CTR partnerships that, among other things,

- Increase the role of scientists and government officials of CTR partner countries in designing CTR programs and projects; and
- Increase financial contributions and additional commitments to CTR programs and projects from Russia and other partner countries, as appropriate, as evidence that the programs and projects reflect national priorities and will be sustainable.

- Include broader international cooperation and partnerships, and increased international contributions;
- Incorporate a strong focus on national programs and sustainability, which includes actions to address concerns raised and recommendations made by the Government Accountability Office, in its report of February 2007 titled "Progress Made in Improving Security at Russian Nuclear Sites, but the Long-Term Sustainability of U.S. Funded Security Upgrades is Uncertain," which pertain to the Department of Defense;
- Continue to focus on the development of CTR programs and projects that secure nuclear weapons; secure and eliminate chemical and biological weapons and weapons-related materials; and eliminate nuclear, chemical, and biological weapons-related delivery vehicles and infrastructure at the source; and
- Include efforts to develop new CTR programs and projects in Russia and the former Soviet Union, and in countries and regions outside the former Soviet Union, as appropriate and in the interest of U.S. national security.

STRUCTURE OF THE REPORT

This report has 5 chapters and 10 appendixes. There is a "Chapter Summary of Findings and Recommendations" at the end of each chapter.

- Chapter 1 summarizes the evolution of CTR activities, beginning with the programs established to respond to the collapse of the Soviet Union through today's broad range of U.S. government and international threat reduction efforts. This chapter is not meant to be encyclopedic, but offers an overview of the program and how it evolved, and highlights some of the many accomplishments that have been achieved under the DOD CTR program.
- Chapter 2 sets out the vision of what CTR 2.0 is and why the program should evolve in that direction.
- Chapter 3 describes the form and function of CTR 2.0, identifying key elements that will characterize this new approach. These elements recognize the need for flexibility and adaptability, the central role played by partnership, and the overarching requirement for clear strategic guidance and leadership from the White House and other senior members of the administration, and new budgetary, legal, and policy tools that are needed. This chapter also provides some examples of the types of activities the committee expects could be undertaken under CTR 2.0.
- Chapter 4 discusses the role that the DOD CTR program might have in CTR 2.0 and provides some illustrations of the types of programs that should be considered. The committee did not prescribe activities for a specific group of countries or region, but rather attempted to demonstrate that by thinking more broadly and creatively about global security engagement, and by engaging

a range of partners, it should be possible to identify meaningful activities for almost any environment of security interest to the United States.

- Chapter 5 addresses strategic implementation issues for CTR 2.0 and how to move from concept to action. The chapter draws together several actions from the findings and recommendations.
- The appendixes provide references and other supporting documentation for the discussions in the report.

INFORMATION SOURCES

The committee members and staff reviewed many relevant reports, some of which were released around the time this report went into review. To the extent possible, the committee considered the findings and recommendations of these and other relevant studies. Key documents are cited in the text, footnotes, and appendixes of the report (see Appendix C).

Additionally, the committee held several meetings in Washington, D.C. (see Appendix D), during which it received briefings from officials and representatives from DTRA, the Departments of Defense, State, Energy, and Health and Human Services, and nongovernmental organizations engaged in implementing and analyzing CTR programs. In response to initial findings, several committee members and staff visited the headquarters of the U.S. Pacific Command, European Command, and the newly formed African Command. Finally, the committee has collected a large library of open-source, publicly available materials on the CTR program to support its research. Following the close of this project, these resources will be made available to the public via the Internet.

The committee and the Department of Defense, as the report sponsor, recognized that discussing options for global security engagement could easily lead to classified issues. Therefore, by mutual agreement, issues such as the role of the intelligence community, the relationship between CTR programs and other security negotiations, and sensitive information on the relationship between the United States and other governments are not explored in this report.

1

The Evolution of Cooperative Threat Reduction

This chapter summarizes the evolution of cooperative threat reduction activities, beginning with the programs established to respond to the collapse of the Soviet Union through today's broad range of U.S. government and international threat reduction efforts (see Appendix E). Just as programs evolved over time, so did the terminology used to refer to those programs. As explained in "A Note on Terminology," this report has adopted the following terms to refer to various programs: programs exclusive to the Department of Defense Cooperative Threat Reduction program are referred to as *DOD CTR*; the broader set of threat reduction programs that encompasses departments and agencies across the U.S. government are referred to as *USG CTR*; the entire set of programs to this point is referred to as *CTR 1.0*; and the committee's concept of a future global security engagement program is referred to as *CTR 2.0*.

DOD CTR has played a central nonproliferation role since 1992, and successfully addressed a myriad of disarmament, dismantlement, and engagement challenges that emerged throughout the 1990s. Senator Sam Nunn and Senator Richard Lugar formally established the program in late 1991, but a history of limited and selective types of interactions between the United States and Soviet Union helped make it possible for the two countries to embark on such a groundbreaking effort. U.S.-Soviet Space Cooperation and the Joint Verification Experiments are two such examples.

A 1985 report by the Congressional Office of Technology Assessment (OTA) looked at the congressional debate on whether to revive the U.S.-Soviet space cooperation that had begun in the 1970s and was allowed to lapse in 1982.[1] Several issues were considered, including scientific and practical benefits

---

[1] U.S. Office of Technical Assessment. 1985. *U.S.-Soviet Cooperation in Space: A Technical Assessment*. 1-3 pp. Availiable as of March 2009 at http://govinfo.library.unt.edu/ota/Ota_4/DATA/1985/8533.PDF.

to be gained, the potential transfer of militarily sensitive technology or know-how, the foreign policy impact of space cooperation, why the Soviet Union wanted to pursue space cooperation, and how all of these issues factored into overall U.S.-Soviet relations. At the time of the report, the United States had a decade of experience with a Soviet relationship that was characterized by OTA as "strained, unpredictable, and ambiguous."[2] However, the report concludes: "From a scientific and practical point of view, past experience has shown that cooperation in space can lead to substantive gains in some areas of space research and applications, and can provide the United States with improved insight into the Soviet space program and Soviet society as a whole."[3] The OTA risk-benefit analysis came out in favor of cooperation.

Similarly, the Joint Verification Experiment Agreement of May 31, 1988, addressed many sensitive nuclear testing issues and ultimately led to an extraordinary set of interactions that allowed scientists, technicians, and observers from the United States and the Soviet Union not only to observe an underground nuclear explosion experiment at each other's test sites, but also to measure explosion yields and discuss the test results.[4] Although clearly distinct from the beginning of the DOD CTR program, these U.S. efforts provided important underpinnings for the DOD CTR effort, especially on the Russian side.[5]

The DOD CTR program has never operated in a vacuum, but rather as a component of much broader national and international efforts. The threats that the United States faces today and is likely to confront in the future are more diverse and complex than were those posed by the former Soviet Union (FSU). The committee has found that DOD CTR and other cooperative threat reduction programs have been successful in the past, and is confident that these programs can be adapted and applied to new situations.

The DOD CTR program was created in response to the unique circumstances surrounding the collapse of the Soviet Union. The events leading to the August 1991 coup and subsequent breakup of the Soviet Union had their roots in the accelerated change inspired by the 1980s *glasnost'* ("openness") policy

---

[2] Ibid. p. 1.

[3] Ibid.

[4] "At a summit in Washington, D.C. in December 1987, the two countries agreed to a set of on-site reciprocal experiments to monitor nuclear explosions at their corresponding test facilities. This culminated in the Joint Verification Experiments (JVE) where Soviet experts monitored a nuclear explosion at the Nevada Test Site on August 17, 1988, and U.S. experts monitored a nuclear explosion at the Semipalatinsk test site on September 14, 1988. . . . The JVEs laid the foundation for future technical cooperation between Russian and American scientists." National Academy of Sciences. 2005. *Monitoring Nuclear Weapons and Nuclear-Explosive Materials: An Assessment of Methods and Capabilities.* Washington, D.C.: The National Academies Press. 31-32 pp. Available as of March 2009 at http://www.nap.edu/catalog.php?record_id=11265.

[5] For a Russian perspective on the contributions of the JVEs, see National Research Council. 2004. *Overcoming Impediments to U.S.-Russian Cooperation on Nuclear Nonproliferation: Report of a Joint Workshop.* Washington, D.C.: The National Academies Press. 71-72 pp.

of Mikhail Gorbachev, but no one could have predicted or planned the way in which events transpired, including those in the vast Soviet military complex. As a result, "All of the [Soviet] military forces were left in place. There were 27,000 nuclear weapons, 40,000 tons of chemical weapons, unknown quantities of biological weapons materials, and 10 closed nuclear cities."[6] Many, but not all, of these military assets were in Russia, but pieces of the Soviet Union's weapons of mass destruction (WMD) systems were distributed across other former Soviet republics. There were increasing fears in the West[7] about the stability of the Newly Independent States (NIS).[8] Diversion of the former Soviet arsenal of WMD and related materials, delivery systems, and expertise later became the primary concern.[9] This was new territory and there were few precedents for how to proceed. But U.S. and Russian officials came to understand that significant rapid action had to be taken to secure the vast arsenals in the NIS. The original U.S. negotiators arrived in Moscow for their first meeting with a blank sheet of paper, ready to listen to proposals from the Russians, who described activities that they felt responded to their highest priorities. Over a series of discussions that were both cooperative and collaborative, an outline of a cooperative threat reduction program began to take shape.

The DOD CTR program was initially authorized in 1991 and supported by funds appropriated to the Department of Defense in Public Law 102-228.[10] The law defined three primary program objectives: (1) assist the former Soviet states to destroy nuclear, chemical, and other weapons; (2) transport, store, disable, and safeguard weapons in connection with their destruction; and (3) establish verifiable safeguards against the proliferation of such weapons. In 1992, these objectives were expanded to include dismantling missiles and missile launchers; destroying destabilizing conventional weapons; preventing diver-

---

[6] Joseph P. Harahan. Discussion at CTR Study Committee Meeting #1. May 21, 2008. See Appendix C for a list of references that address the history of the CTR program.

[7] In November 1991, the Carnegie Corporation of New York convened a meeting to address the Soviet nuclear arsenal. After briefings from Ashton Carter and William Perry, Senator Sam Nunn, Senator Richard Lugar, and their senior staff worked together to draft legislation that passed the Senate later that month. See Ashton B. Carter and William J. Perry. 1999. *Preventive Defense: A New Security Strategy for America.* Washington, D.C.: Brookings Institution Press. 71-72 pp.

[8] The Newly Independent States (NIS) refer to the countries formed on the basis of the former Soviet Republics, and does not include the Baltic States.

[9] See for example Graham Allison et al., eds. 1993. *Cooperative Denuclearization: From Pledges to Deeds.* CSIA Studies in International Security No. 2. Harvard Project on Cooperative Denuclearization. Center for Science and International Affairs: Harvard University.

[10] Public Law 102-228 (section 2551 [note], title 22, United States Code), Soviet Nuclear Threat Reduction Act of 1991, December 12, 1991. Congress initially authorized the transfer of $400 million in each of FY 1992 and FY 1993 for CTR activities under Section 108 of the "Dire Emergency Supplemental Appropriations and Transfers for Relief from the Effects of Natural Disasters, for Other Urgent Needs, and for Incremental Cost of Operation Desert Shield/Desert Storm Act of FY 1992," P.L. 102-228, as amended and Section 9110(a) of the National Defense Appropriations Act for FY 1993, P.L. 102-396.

TABLE 1.1 DOD CTR Funding: Requests and Authorization ($ millions)

Fiscal Year (FY)	1992	1993	1994	1995	1996	1997	1998
Request	$400	$400	$400	$400	$371	$328	$382.2
Authorized	$400	$400	$400	$400	$300	$364.9	$382.2
Fiscal Year	**1999**	**2000**	**2001**	**2002**	**2003**	**2004**	**2005**
Request	$440.4	$475.5	$458.4	$403	$416.7	$450.8	$409.2
Authorized	$440.4	$475.5	$443.4	$403	$416.7	$450.8	$409.2
Fiscal Year	**2006**	**2007**	**2008**				
Request	$415.5	$372.3	$348.00	Total Requested FY 1992-2008			$6,870.70
Authorized	$415.5	$372.3	$428.05	Total Authorized FY 1992-2007			$6,901.85

SOURCE: Amy Woolf. 2008. *Nonproliferation and Threat Reduction Assistance: U.S. Programs in the Former Soviet Union.* Washington, D.C.: Congressional Research Service. 11 pp. Available as of March 2009 at http://fas.org/sgp/crs/nuke/RL31957.pdf.

sion of weapons-related scientific expertise; establishing science and technology centers; facilitating demilitarization of defense industries and converting military capabilities and technologies; and expanding military-to-military and defense contacts.[11]

After a slow start-up process in 1992 and 1993, some individuals in Congress criticized the DOD CTR program for spending too much time and money on what were considered "soft" activities as opposed to the "hard," more tangible, WMD dismantlement and destruction programs. In response to congressional preferences, some programs originally established and funded by the Department of Defense, as explained below, are now funded by the Departments of State and Energy. Other programs, such as Defense Conversion (investment assistance to convert former Soviet military infrastructure to peaceful, civilian, commercial purposes) and Military Officer Housing (to accelerate the retirement of former Soviet military officers) lost congressional support and were eliminated altogether.

The United States has invested more than $21 billion in USG CTR programs since 1992, nearly one-third of which was for DOD CTR. See Table 1.1 for a summary of DOD CTR funding over the life of the program.

Despite some difficulties over the years, the DOD CTR funding has accomplished a great deal in the region to increase security and prevent the potential diversion of weapons of mass destruction and associated technologies, materials, and expertise. As of February 2009, the United States and the NIS have deactivated 7,504 strategic nuclear warheads, destroyed 742 interconti-

---

[11] 1993 National Defense Authorization Act, Public Law 102-484, October 23, 1992, Title XIV – Demilitarization of the Former Soviet Union (also cited as the "Former Soviet Union Demilitarization Act of 1992").

nental ballistic missiles (ICBMs), eliminated 496 ICBM silos, destroyed 143 ICBM mobile launchers, eliminated 633 submarine-launched ballistic missiles (SLBMs), eliminated 476 SLBM launchers, destroyed 31 nuclear submarines, and launched biological surveillance efforts in several NIS states.[12]

Finding 1-1: The DOD CTR programs have demonstrated that DOD was able to mobilize and focus considerable resources creatively to meet new challenges in Russia and other states of the former Soviet Union. In particular, DOD showed that it could apply assistance to deactivate nuclear warheads, eliminate chemical munitions, delivery systems, and biological and chemical production facilities in a verifiable and transparent way.

DOD CTR OPERATES IN TANDEM WITH OTHER U.S. PROGRAMS

From the outset, DOD CTR program execution depended heavily on the diplomatic leadership of the Department of State and the nuclear weapons expertise of the Department of Energy (DOE). Both departments played active roles in the process of negotiating the DOD CTR Umbrella and Implementing Agreements, provided expertise, and initially received funds from DOD for program implementation. Although authorized, funded, and identified initially as a DOD program, over time, the concept of cooperative threat reduction has grown into an interagency enterprise that encompasses the resources and expertise of many U.S. government departments and agencies, including several that have not traditionally had a national security role.

In 1996, responding to congressional criticism of the DOD CTR program, a decision was reached among the secretaries of defense, state, and energy to transfer funding responsibility for certain activities out of the Defense Department budget request to the State and Energy Department budgets. The State Department became responsible for annual appropriations requests for the WMD Scientist Redirection Program and the Export Controls and Border Security Program, and continued to fund the Nonproliferation and Disarmament Fund (NDF) under its existing FREEDOM Support Act authorities.[13] The Department of Energy took responsibility for programs of Material Protection, Control and Accounting to protect, secure, and account for nuclear materials and for a new program aimed at facilitating the transformation and downsizing of Russia's large nuclear research and fissile material production facilities. Further devolution of program funding responsibility resulted from the George W. Bush administration's Review of Nonproliferation Assistance to Russia,

---

[12] See Appendix F for the most recent Nunn-Lugar Scorecard. Available as of March 2009 at http://lugar.senate.gov/nunnlugar/scorecard.html.

[13] See The FREEDOM Support Act, P.L. 102-511. October 24, 1992. Available as of March 2009 at http://www.fas.org/nuke/control/ctr/docs/s2532.html.

completed in December 2001. That review concluded that most programs were effective and well run, and made several recommendations that were reflected in the Fiscal Year 2002 budget requests to Congress. In particular, this included the transfer of $74 million in funding as well as future funding responsibility for the project to Eliminate Weapons-Grade Plutonium Production to DOE. The one-time transfer of $30 million to the Department of State to fund Biological Weapons Redirection efforts was also recommended.[14]

Over time, other departments not typically considered to have a national security function were enlisted to support these efforts, particularly for programs administered by the State Department. These include the Departments of Health and Human Services and Agriculture, and the Environmental Protection Agency, which support the WMD Scientist Redirection programs, and the Departments of Treasury and Commerce, the U.S. Customs Service, and U.S. Coast Guard, which support implementation of the Export Controls and Border Security Program. All of them brought scientific, technical, training, and other expertise necessary for program implementation and oversight that were not available elsewhere in the government.

It has also become apparent that development assistance, such as that provided through the U.S. Agency for International Development (USAID), also can play a role to support cooperative threat reduction efforts. In some areas of interest to USG CTR, such as projects related to disease monitoring and health, USAID's programs and budgets can be leveraged to complement and supplement USG CTR efforts of other agencies, and often have much larger budgets. In addition, private foundations are now major players in funding a wide variety of programs that can operate synergistically with threat reduction programs. For example, the health investments of the Bill and Melinda Gates Foundation, Google.org, and others are examples of programs that could work with CTR 2.0 efforts in disease surveillance and biological threats. Similarly, the Nuclear Threat Initiative and Global Green have worked with both governments and international organizations to address nuclear and chemical security challenges.

THE KANANASKIS G8 SUMMIT AND THE CREATION OF THE GLOBAL PARTNERSHIP AGAINST THE SPREAD OF WEAPONS AND MATERIALS OF MASS DESTRUCTION

In addition to expanding U.S. interagency involvement, USG CTR concepts also were firmly incorporated into the Group of Eight (G8) agenda at the 2002 Kananaskis Summit, where leaders created the *G8 Global Partnership*

---

[14] The White House, Office of the Press Secretary. United States Government Nonproliferation/Threat Reduction Assistance to Russia Fact Sheet May 24, 2002. Available as of March 2009 at http://www.fas.org/nuke/control/sort/fs-nonpr.html.

(G8 GP) Against the Spread of Weapons and Materials of Mass Destruction.[15] The G8 GP defined its mission as preventing "terrorists, or those that harbor them, from acquiring or developing nuclear, chemical, radiological, and biological weapons; missiles; and related materials, equipment and technology."[16] Programs were implemented initially in Russia, but later other countries of the FSU also participated. The G8 GP was a response to the September 11, 2001, terrorist attacks and the October 2001 anthrax mailings. It defined its international efforts in terms of preventing weapons and materials of mass destruction acquisition by terrorists rather than on more traditional state-supported programs. The 2007 G8 GP Mid-Point Review reflected progress in neutralizing and destroying Russian chemical weapons, dismantling decommissioned nuclear submarines, disposing of fissile materials, employing former WMD scientists and engineers in civil activities, and enhancing the safety of nuclear materials. (See Box 1.1 below for details on the program to Eliminate Chemical Weapons in Russia.)

The development of the G8 GP elevated cooperative threat reduction to a global enterprise that is now poised to extend beyond its original 10-year mandate. Common guidelines[17] have been established for program implementation and informal mechanisms have proven effective as a relatively low-cost, low-bureaucracy mode of program coordination. For example, ad hoc technical coordination groups for chemical weapons destruction and submarine dismantlement projects meet only when necessary and conduct much of their business through e-mail, conference calls, and other similar means.

At the 2008 G8 Summit in Japan, the G8 GP Report noted that "We also recognize that the GP must evolve further to address new, emerging risks worldwide if we are to prevent terrorists or those that harbor them from acquiring chemical, biological, radiological, nuclear weapons and/or missiles,"[18] and elaborates further on how the GP would be expanded, noting that 23 countries now contribute to GP efforts and that more should be encouraged to join.[19] (See Box 1.2.)

---

[15] G8 2002 Kananaskis Summit Agenda. Available as of March 2009 at http://www.canadainternational.gc.ca/g8/summit-sommet/2002/index.aspx?menu_id=15&menu=L.

[16] G8 Leaders. 2002. Statement at Kananaskis Summit: The G8 Global Partnership Against the Spread of Weapons and Materials of Mass Destruction. June 27. Available as of March 2009 at http://www.canadainternational.gc.ca/g8/summit-sommet/2002/global_partnership-partenariat_mondial.aspx?lang=eng. See also Charles Thornton. 2002. The G8 Global Partnership Against the Spread of Weapons and Materials of Mass Destruction. *The Nonproliferation Review.* 9:3.

[17] See Appendix G for the G8 GP Guidelines for New and Expanded Cooperation Projects.

[18] G8 Countries. 2008. Report on the G8 Global Partnership. Hokkaido Toyako Summit. Available as of March 2009 at http://www.mofa.go.jp/policy/economy/summit/2008/doc/pdf/0708_12_en.pdf.

[19] The participants in the G8 GP include the G8: Canada, France, Germany, Italy, Japan, the Russian Federation, the United Kingdom, the United States, as well as Australia, Belgium, the

> **BOX 1.1**
> **G8 Global Partnership Efforts to Eliminate Chemical Weapons in Russia**
>
> In the 2008 Report on the Global Partnership, the following progress was highlighted in the Russian Chemical Weapons Destruction project since 2002, noting that international contributions to the project include funding from the government of Russia:
>
> - Two chemical weapons destruction facilities were built:
>   o **Gorny**
>   - Assistance is provided from the European Union (EU), Finland, Germany, the Netherlands, and Poland
>   - All chemical weapons stored there have been neutralized
>   o **Kambarka**
>   - Assistance is provided from the EU, Finland, Germany, the Netherlands, Sweden, and Switzerland
>   - The facility became operational in December 2005, and has been neutralizing chemical weapon stockpiles since
> - The facility at **Shchuch'ye** is being constructed:
>   o Assistance provided from Belgium, Canada, the Czech Republic, the EU, France, Ireland, Italy, the Netherlands, New Zealand, Norway, Sweden, Switzerland, the United Kingdom (UK), and the United States
>   o Finland also plans to contribute to this project
> - Additional assistance will be provided to the facilities at
>   o **Pochep**
>   - Assistance received from Germany and Switzerland
>   - Italy also plans to support this project
>   o **Leonidovka** and **Maradykovsky**
>   - Switzerland has provided assistance to both sites
>   o **Kizner**
>   - Canada is preparing to provide assistance
>
> SOURCE: Report on the G8 Global Partnership. 2002. Paragraphs 29-32. Available as of March 2009 at http://www.mofa.go.jp/policy/economy/summit/2008/doc/pdf/0708_12_en.pdf.

Finding 1-2: The DOD CTR program in Russia and the former Soviet Union is a vital part of the broader interagency and international cooperative threat reduction efforts, and operates in the context of a broader group of U.S. interagency and international programs.

Czech Republic, Denmark, the European Union, Finland, Ireland, the Netherlands, New Zealand, Norway, the Republic of Korea, Sweden, and Switzerland.

BOX 1.2
Expansion Of The G8 Global Partnership

29. Risks of the spread of weapons and materials of mass destruction exist worldwide. The Global Partnership (GP) will address such risks through implementing projects according to the GP common principles. In addressing threat reduction and non-proliferation requirements, the projects will be specifically aimed to implement and realize the GP common principles worldwide. To this end other recipient states and donor states accepting the GP principles and guidelines could be included on a case-by-case basis in an expanded GP for the implementation of projects in line with GP goals. At the same time, the GP will continue to focus on the ongoing GP projects.

30. At the same time, the GP will continue to provide assistance to ongoing GP projects in Russia noting that the areas of the chemical weapons destruction and the dismantlement of decommissioned nuclear submarines are priority areas for Russia. We are determined in our commitment to accomplish projects, including those which Russia considers of primary importance, under this initiative in Russia.

31. Based on the agreement that the Global Partnership will address such risks worldwide, the partners will work together constructively and practically to identify specific focuses of the expanded GP. The discussions on this issue will be conducted on a project based fashion and function-wise, inter alia, nuclear and radiological issues, chemical issues and biological issues. The GP welcomes the expertise of the [Organization for the Prohibition of Chemical Weapons] OPCW on chemical issues and the [International Atomic Energy Agency] IAEA on nuclear and radiological issues in the implementation of GP projects in their area of competence and seeks such expertise regarding biological issues within the Biological and Toxin Weapons Convention (BTWC). The effective implementation of IAEA safeguards agreement and the Additional Protocol, UN Security Council Resolution 1540 and the Global Initiative to Combat Nuclear Terrorism are areas where partners may seek to engage through the GP. A "model agreement" proposed by the UK was noted as a reference which could be helpful in enabling new projects to be put in place with minimum delay.

32. The Global Partnership currently encompasses twenty-three partners including the EU. Efforts should, however, continue to be made to find new donors. Endeavors to communicate with potential new donors can be undertaken by interested partners.

SOURCE: Report on the G8 Global Partnership. 2008 Paragraphs 29-32. Available as of March 2009 at http://www.mofa.go.jp/policy/economy/summit/2008/doc/pdf/0708_12_en.pdf.

DOD CTR INITIAL PROGRAM CHARACTERISTICS

The DOD CTR program had few precedents to guide its initial development, but there was a sense of urgency that drove the first set of activities aimed at consolidating the former Soviet nuclear capabilities that were spread across four of the NIS (Belarus, Kazakhstan, Russia, and Ukraine). That urgency was shared by leaders in both Russia and the United States, in some cases for different reasons. Russia's new leaders were interested in remaining the sole nuclear power in the region, but also recognized that foreign financial assistance would be critical to consolidate, safeguard, and in some cases dismantle weapons systems as well as to help the country through a turbulent economic period.[20] U.S. leaders were concerned about the potential threat from four new nuclear states and about accountability for any U.S. assistance provided for threat reduction and how to ensure that assistance provided was not used to sustain or enhance former Soviet weapons capabilities.

DOD policies, procedures, and rules developed to implement its CTR program were complex, and the process of putting agreements into place to govern the new program activities were unfamiliar to the leaders of the NIS. In the United States, some individuals in Congress were unconvinced that the USG CTR programs were in U.S. national security interests and saw the program more as foreign assistance.[21] There was still distrust and fear of Russian motives. That unease eventually resulted in very intense oversight of the program and restrictions placed on the types of activities that could be implemented. Auditing and accounting practices, limitations on liability, access to sensitive sites, and other factors became the subject of often lengthy negotiations. Congress has consistently maintained close oversight particularly over the DOD CTR program. Many DOD CTR programs have changed over the years, often in response to congressional directions, restrictions, prohibitions, and preferences. The original legislative mandate for the program required, among other things, a lengthy annual certification measuring against six criteria (including human rights). Oversight was also exercised through more than 40 congressionally requested Government Accountability Office reports on program activities, and there was a general sense of caution that came from wanting to avoid any appearance of programs contributing to helping Russia expand Soviet-era military power. These layers of oversight may have provided an increased sense of political and management security, but also resulted in a heavy bureaucratic burden and implementation delays. The challenge of demonstrating the national security benefits of CTR 2.0 will require an ongoing set

---

[20] Joseph P. Harahan. 2008. Discussion at CTR Study Committee Meeting #1. May 21.
[21] See Ashton B. Carter and William J. Perry. 1999. *Preventive Defense: A New Security Strategy for America.* Washington, D.C.: Brookings Institution Press. 74-75 pp. See also Richard Soll. 1995. Misconceptions About the Cooperative Threat Reduction Program. *Director's Series on Proliferation, 8.* Lawrence Livermore National Laboratory, the University of California.

of consultations between the executive and legislative branches to ensure that members of Congress and their staffs understand the program's strategy and approaches.

Since 1995, the level of leadership of DOD CTR has been downgraded from a high-priority program managed by a deputy assistant secretary of defense for cooperative threat reduction and special assistant to the secretary of defense to a CTR Policy Office under a director for the CTR program.[22] Historically, DOD CTR has been very effective when it had the active and direct support and participation of the secretary of defense. This kind of sustained, senior-level support will be needed in the future.

> Since 1995, the level of leadership in DOD has been downgraded from a high-priority program managed by a deputy assistant secretary of defense for cooperative threat reduction and special assistant to the secretary of defense to a CTR Policy Office under a director for the CTR program.

As DOD CTR grew through the 1990s, there was little corresponding growth in the size of the DOD CTR Policy Office staff that provided overall policy and program guidance. A small and dedicated policy team was expected to provide guidance and policy oversight for a burgeoning number of projects under the supervision of the DTRA CTR Implementing Office. In addition, the programs spread into Central Asia and the Caucasus regions, with each new country requiring an investment of time to establish new working relationships, which were primarily the responsibility of the policy staff. DOD and DOE CTR 1.0 programs benefited from having either DTRA or DOE staff at embassies in countries where programs were implemented. This in-country liaison and oversight function is important to the program, but will be harder to sustain as global expansion of programs puts even more demand on the limited number of staff.

The use of large American contractors with experience working in international environments and with DOD procurement rules initially was a key to successful CTR program implementation. These contractors took on the responsibility of integrating themselves and foreign subcontractors with local firms, and basically became the face of DOD CTR in countries across the FSU. Yet, this growing reliance on contractors created greater separation between DOD CTR policy leaders and their counterparts in cooperating countries, weakening their development of close working relationships and undermining a primary benefit of the early DOD CTR program. The committee was told that for a period of time integrated program reviews were held quarterly for some programs that brought together DOD CTR officials, U.S. contractors, foreign

---

[22] See Carter and Perry. pp. 72-73.

contractors, and subcontractors, including nongovernment organizations. This approach apparently worked well, but was not applied uniformly across all programs.

Finding 1-3: The size of the DOD CTR Policy Office staff has not expanded significantly over the life of the program even though the number of countries engaged has continued to grow, and it will need to expand further to meet the increased requirements of global engagement.

Other transitions took place over time within the DOD CTR program as well as in the broader USG CTR efforts. Although the threats presented by military hardware were still important, the U.S. experience with terrorism demonstrated how important reaching the softer components of threats to security was, and more attention was focused on this area. For example, as the DOD CTR program focus moved from Russia into Central Asia and the Caucasus areas, there was less WMD equipment and infrastructure destruction and dismantlement work. Although there were WMD infrastructure and facilities to address, the programs shifted more to training personnel for security, protecting and securing highly dangerous pathogens, and preventing the movement of WMD materials across insecure borders. Similarly, other USG CTR programs added radiological security and security of highly toxic chemicals to their program portfolios. There was also a shift in how "threats" were defined. In the early years of the DOD CTR program, the emphasis was on WMD threats, particularly strategic weapons systems. Over time, policy makers came to understand that not only those with direct, past weapons experience pose a risk, but also those *capable* of creating weapons threats pose a risk and should be included in programs. These trends are good indicators of how the programs can evolve further in the future to address new threats.

How to expand into some of these areas was not always well thought out, however. For example, the DOD CTR Threat Agent Detection and Response Program was designed to secure repositories of especially dangerous pathogens, enhance surveillance and response to disease outbreaks, and enhance local diagnostics capabilities. As documented in the 2007 study *The Biological Threat Reduction Program of the Department of Defense*,[23] there were insufficient local consultations when developing the program, with the result that the program responded more to DOD CTR "wants" than any local "needs" that would help ensure sustainability. DOD CTR was eventually convinced to modify the list of pathogens that the program will monitor from only those on the U.S. list of

---

[23] National Research Council. 2007. *The Biological Threat Reduction Program of the Department of Defense: From Foreign Assistance to Sustainable Partnerships.* Washington, D.C.: The National Academies Press. Available as of March 2009 at http://www.nap.edu/catalog.php?record_id=12005.

select agents to a somewhat broader list that included endemic disease of real relevance to the partner country. As a result, the partner attitude toward the project improved and the likelihood for sustaining the program into the future increased because of the higher level of local interest and commitment. The issue of data reporting, however, remains unresolved. The original plan was to transmit all data to a DOD end point in the United States, with no plan for how to then share the information with the World Health Organization. The host country would not agree to this plan, and an appropriate nonmilitary host for the data is being identified.

Finding 1-4: The selection of activities in countries with which we engaged in CTR 1.0 was not always done with long-term strategic thought or appropriate awareness of country and regional concerns.

From the Russian perspective, the early period of the DOD CTR program was dominated by the Russian presidential administration and powerful ministries (such as the Ministries of Defense and Atomic Energy), which strongly preferred operating in the context of legal frameworks and implementing agreements. There was strong Russian motivation to implement the program by several key military and nuclear complex leaders who shared U.S. concerns about treaty compliance and meeting treaty compliance milestones and nuclear security. In the early bilateral negotiations, the United States was able to obtain Russian agreement on virtually all of its many implementing requirements and procedures. Later, as Russia stabilized and grew wealthier, and particularly under the Vladimir V. Putin administration, the implementation environment became increasingly challenging. Guidelines were included in the G8 GP agreement at the 2002 Kananaskis Summit[24] that reflected the difficulties that individual states had in winning Russian agreement to certain project elements and overall Russian reluctance to work with other countries under the same conditions that were required by the United States. Access to facilities became increasingly problematic; liability for the actions of foreign contractors working in Russia became a "show-stopping" issue that took several years to resolve. Implementation roadblocks became a regular discussion item at G8 GP meetings. Yet despite the bureaucratic challenges, there are still many *tangible* examples of DOD CTR accomplishments, as demonstrated on the Nunn-Lugar Scorecard[25] and other assessments of USG CTR programs.

The USG CTR programs also have produced equally important *intangible* benefits. The human relationships that have been formed at multiple levels are among the most important, enduring, and underrecognized benefits of these programs. Working from a basis of shared priorities, strategies, goals, and

---

[24] See Appendix G for the G8 GP Guidelines for New and Expanded Cooperation Projects.
[25] See Appendix F.

responsibilities in a truly cooperative environment produces more than just tangible program success. The concept of long-term engagement, the development of lasting ties based on trust built through shared experience, defies the hard metrics that have become such an ingrained part of measuring program value, but can be the critical link to success of an immediate project and, perhaps more importantly, be the foundation for working together in future endeavors. These links have been major contributors to success in the former Soviet Union, and time and effort must be invested in each new environment to develop these relationships. Perhaps most importantly, these relationships have helped the United States gain insights into personalities and government structures that make it possible to design more effective approaches to cooperation.[26] This is true not only for the partner or recipient countries but also for the countries with which the United States collaborates through the G8 GP and other international or multilateral structures.

Finding 1-5: DOD CTR is a highly leveraged national security program for the United States that yields reciprocal insights and transparency that can lead to greater levels of trust and confidence.

DOD CTR AT AN INFLECTION POINT

As most DOD CTR activities in Russia move toward completion and as security threats beyond the FSU become a new priority, the DOD CTR program finds itself at an inflection point. In 2002, Congress began asking DOD to explore ways that the program could be used to meet new global challenges.[27] The DOD CTR authorizing legislation was changed in 2003 to allow activities outside the FSU and shortly thereafter a project was developed to help Albania destroy a chemical weapons cache left from the Cold War period.[28] The initial scoping study for the project was undertaken by the State Department (NDF), and DOD CTR's chemical weapons stockpile destruction assistance to Albania began in 2006, at a facility designed under DOD CTR supervision. The project was completed in 2007, and remains the only DOD CTR project undertaken outside the territory of the FSU.[29]

---

[26] Charles Thornton. 2008. Discussion at the CTR Study Committee Meeting #1, May 21.

[27] In 2002, Congress requested a report outlining a cooperative threat reduction program for India and Pakistan, including legal obstacles to implementing such a program, and an estimated budget. The report was apparently never produced and DOD could not provide any documentation about the report to the committee.

[28] Brianne E. Tinsley. Defense Threat Reduction Agency. Albania Chemical Weapons Elimination Project. Project Overview. Breifing. Department of Defense. October 14, 2008.

[29] "Project Peace" to eliminate the former Soviet Large Phased Array Radar (LPAR) at Skrunda, Latvia, was funded under CTR in 1994-1995 from a $10 million earmark of CTR funds for conventional weapons dismantlement. The elimination of the LPAR was a primary obstacle to Russian

Another opportunity for DOD CTR program expansion was in response to Libya's announcement in December 2003 that it was giving up its WMD programs. Among the WMD elements almost ideal for DOD CTR program involvement were providing the transportation for the removal of gas centrifuges and nuclear material and destroying Libya's chemical weapons, precursor chemicals, and related manufacturing capability. However, the DOD CTR program estimate of time, cost, and complexity significantly exceeded the estimates from the State Department's NDF, which could operate with greater flexibility given its "notwithstanding any other provision of law" authority (which allows it to operate legally in any environment, even when sanctions or other measures may be in place). Although NDF ultimately did not implement the Libyan chemical weapons destruction program, the committee studied the Libya case to understand better why DOD CTR was not involved.[30]

Future DOD CTR and other USG CTR programs may be similarly affected by evolving political and economic relationships with partner nations. Tensions with Russia after its August 2008 conflict with Georgia raised questions at the leadership level for USG CTR implementing agencies, but at the program level there was little impact. Congressional committee staff with direct interests in USG CTR efforts were uniform in their support for sustaining USG CTR programs, including DOD CTR, despite growing tensions with Russia.[31] As the global security environment continues to evolve, there will be times when the United States, Russia, the participants in the G8 GP, or others may be at odds over objectives or courses of action for issues with no direct relationship to cooperative threat reduction. It would be a great loss for U.S. and international security if temporary political turmoil were to have a negative impact on the long-term efforts under the DOD CTR program, USG CTR programs, or G8 GP efforts. Stepping away from programs in Russia would have risked sacrificing many gains that have been made in the past decade and a half and which, once lost, might never be regained. This is not an option when Russia and other countries are partners in important global security efforts, such as the denuclearization of the Democratic People's Republic of Korea.

In the course of the study that led to this report, the committee explored how the CTR concept can be applied to contemporary threats. When the CTR programs were conceived, they were intended to address the primarily monolithic problem of the Soviet Union's WMD capacity and related proliferation risks. Although a diverse and complex set of challenges, those issues

---

and Latvian agreement on the removal of Russian troops from Latvian territory. Although the Baltic States of Estonia, Latvia, and Lithuania were annexed by the Soviet Union in 1945, this annexation was never officially recognized by the United States, which continued to consider the Baltic States as independent nations.

[30] See further discussion on Libya in Chapter 2.

[31] Communications with Senate and House Armed Services Committees staff and Senate Foreign Relations Committee staff.

were largely concrete and identifiable. The threats that the world confronts in the 21st century, however, are of a fundamentally different nature. Because of rapid globalization of communications, transportation, and knowledge, threats are networked, agile, adaptable, and difficult to quantify; our tools to respond to this kind of threat must be similarly nimble. In the committee's view, a new approach to CTR is now required.

Finding 1-6: The DOD CTR program will require new energy and creativity to deal with the changing global security environment, whose challenges are different from those that came at the end of the Cold War.

The world has changed enormously since the DOD CTR program was established. The events of September and October 2001 triggered a fundamental rethinking of how the United States defines threats and how to respond to them. WMD proliferation focus began to shift from destroying weapons and materials and preventing the flow of expertise and technology from state programs to preventing terrorist acquisition of WMD. Threats from the dual-use potential of known and emerging technologies also had to be managed, as well as the potential for the diversion of industrial chemical or biological materials to malevolent use.

Another new challenge was the ability of an individual or group to cause enormous damage, disruption, and economic loss to the United States, even without widespread death or illness. As demonstrated by the terrorist and biological attacks of 2001, neither massive Soviet-style weapons production facilities nor ICBMs loaded with biological or nuclear payloads are needed to have a significant impact on our society. In addition to the potential for nonstate actors to pose significant threats, many states now have the latent scientific and technical capability to move rapidly into WMD development. Policies should aim at preventing this, but conversely, we must also be prepared to respond positively to countries that may decide to relinquish their weapons programs. The case of Libya demonstrated that such decisions can be made somewhat abruptly and that the United States and other nations require program flexibility to be ready to respond to such unanticipated opportunities and to sustain that response.

Finding 1-7: Most future threats to the United States are likely to have smaller footprints, less distinct signatures, and be more closely associated with industrial activities related to energy, biology, health, or chemistry rather than highly centralized, large-scale national weapons programs.

Recommendation 1-1: The DOD CTR program should be expanded geographically, updated in form and function according to the concept proposed in this report, and supported as an active tool of foreign policy by engaged leadership from the White House and the relevant cabinet secretaries.

CHAPTER SUMMARY OF FINDINGS AND RECOMMENDATIONS

Finding 1-1: The DOD CTR programs have demonstrated that DOD was able to mobilize and focus considerable resources creatively to meet new challenges in Russia and other states of the former Soviet Union. In particular, DOD showed that it could apply assistance to deactivate nuclear warheads, eliminate chemical munitions, delivery systems, and biological and chemical production facilities in a verifiable and transparent way.

Finding 1-2: The DOD CTR program in Russia and the former Soviet Union is a vital part of the broader interagency and international cooperative threat reduction efforts, and operates in the context of a broader group of U.S. interagency and international programs.

Finding 1-3: The size of the DOD CTR Policy Office staff has not expanded significantly over the life of the program even though the number of countries engaged has continued to grow, and it will need to expand further to meet the increased requirements of global engagement.

Finding 1-4: The selection of activities in countries with which we engaged in CTR 1.0 was not always done with long-term strategic thought or appropriate awareness of country and regional concerns.

Finding 1-5: DOD CTR is a highly leveraged national security program for the United States that yields reciprocal insights and transparency that can lead to greater levels of trust and confidence.

Finding 1-6: The DOD CTR program will require new energy and creativity to deal with the changing global security environment, whose challenges are different from those that came at the end of the Cold War.

Finding 1-7: Most future threats to the United States are likely to have smaller footprints, less distinct signatures, and be more closely associated with industrial activities related to energy, biology, health, or chemistry rather than highly centralized, large-scale national weapons programs.

Recommendation 1-1: The DOD CTR program should be expanded geographically, updated in form and function according to the concept proposed in this report, and supported as an active tool of foreign policy by engaged leadership from the White House and the relevant cabinet secretaries.

2

Cooperative Threat Reduction in the 21st Century: Objectives, Opportunities, and Lessons

CTR 2.0 – FROM PATCHWORK PROGRAMS TO HOLISTIC APPROACH

> As the change in version number indicates, CTR 2.0 is a major program upgrade, not just a set of minor patches.

Cooperative Threat Reduction (CTR) 1.0 has matured and developed, and the time has come to move beyond an ad hoc collection of weapons of mass destruction (WMD) nonproliferation activities in the former Soviet states. That set of programs was a highly creative response to unique security challenges and geopolitical changes, particularly in the former Soviet Union (FSU). A new, equally creative set of integrated and coordinated global security engagement programs is now required to address a broader range of WMD and terrorist threats on a global scale—CTR 2.0. As the change in version number indicates, CTR 2.0 is a major program upgrade, not just a set of minor patches. The committee envisions a version of CTR that builds on a proven platform and the lessons learned from the FSU experience, but with substantially new features. As CTR 2.0 grows, it will absorb the lessons learned from the original programs, and will be structured to respond to a rapidly changing environment. CTR 2.0 will not be the domain of a single U.S. department or even of the U.S. government, and the White House will need to play an active leadership role. To succeed, it will need to be an integrated, cooperative, collaborative, global enterprise that is responsive, flexible, and adaptable (see Box 2.1).

> **BOX 2.1**
> **Defining CTR 2.0**
>
> CTR 2.0 is a set of programs and projects undertaken by the United States, as part of a cooperative network that includes a wide range of countries, international organizations, and nongovernment partners, to prevent, reduce, mitigate, or eliminate common threats to U.S. national security and global stability that have emerged in particular since the end of the Cold War. The preferred mechanism and long-term goal for the cooperation is partnership, which means that the countries participating should be ready to share responsibilities for project definition, organization, management, and financing according to a rational division of labor, capacity (including budget capacity), or technical capability. Although CTR 2.0 engagements may have to begin under less than ideal circumstances, the goal for countries engaged under CTR 2.0 is shared responsibility through engagement and partnership. CTR 2.0 should be capable of rapid response as well as longer-term programmatic engagement.

DEFINING CTR 2.0

In late 2003, Libya agreed to give up its WMD programs and join or rejoin relevant international institutions. Although the announcement was preceded by talks between Libyan, U.S., and U.K. government officials, Libya's decision surprised many other than the few individuals directly involved in the negotiations. Weapons, materials, and systems needed to be removed quickly and with a high level of international coordination. U.S. and U.K. officials feared that the Libyans might reverse their decision, and a rapid and flexible response was needed.

Nuclear dismantlement required a creative partnership between Libyan, U.S., U.K., and Russian officials, the U.S. Department of Energy's (DOE) National Nuclear Security Administration (NNSA), the International Atomic Energy Agency (IAEA), and the Nuclear Threat Initiative (NTI) (a U.S. nongovernmental organization [NGO]).[1] The Libyan government agreed to dismantle its centrifuge program and convert its research reactor core from highly enriched uranium (HEU) to low-enriched uranium. Sensitive centrifuge equipment and both spent and fresh HEU fuel were removed under NNSA's Global Threat Reduction Initiative. The Department of Defense (DOD) was asked to provide air or sea transportation for this cargo. Instead, transportation was secured by the State Department and funded by its Nonproliferation and

---

[1] Paula DeSutter. 2004. Completion of Verification Work in Libya. Testimony of Assistant Secretary of State for Verification and Compliance before the Subcommittee on International Terrorism, Nonproliferation, and Human Rights, September 22. Available as of March 2009 at http://www.state.gov/s/l/2004/78305.htm.

Disarmament Fund (NDF), which leased aircraft and a cargo ship. Senior U.S. government officials that were directly involved in the discussions informed the committee that DOD claimed that wartime constraints on military aircraft made it impossible to provide the capability that was needed to remove the equipment and material.

Libya also agreed to destroy their chemical weapons stocks and dismantle their production capability.[2] More than 23 tons of mustard blister agent along with 600 tons of precursor chemicals had to be destroyed.[3] The DOD CTR program and NDF were both asked to submit time and cost estimates. NDF proposed a significantly lower budget and a shorter time line, and it was therefore selected to carry out the task. NDF subsequently contracted with an Italian company to build an incinerator to destroy the chemical weapons material.[4]

Large infrastructure projects were hallmarks of DOD CTR in the FSU and responded to the needs of the massive Soviet military infrastructure–warhead storage facilities, chemical destruction plants, and replacement power plants for plutonium production reactors. A similar situation is likely to be rare in the future. None of the nuclear or chemical engagement in Libya[5] involved DOD CTR in large measure because the existing program is not structured to respond quickly and the senior DOD officials who can direct resources are not sufficiently engaged. The Libyan example shows the need for and importance of a robust, fast, and flexible U.S. government (USG) CTR capability to meet new challenges when they arise.

Finding 2-1: CTR 1.0 was a highly creative response to unique security challenges and geopolitical changes in the former Soviet Union. The new threats we face require similar innovation to create CTR 2.0. Coordination and leadership from the White House will be required, and relevant departments and agencies will need to engage to ensure that there is a clear connection between the policy intent and program implementation, as in the case of Libya.

---

[2] Joseph Cirincione et al. 2005. *Deadly Arsenals: Nuclear Biological and Chemial Threats: Second Edition.* Washington, D.C.: Carnegie Endowement for International Peace. 324 pp.

[3] Ambassador Donald Mahley. Committee consultation. See also Donald Mahley. 2004. Dismantling Libyan Weapons: Lessons Learned. *The Arena*. Chemical And Biological Arms Control Institute. November.

[4] Libya, the United States, the United Kingdom, and the Organization for the Prohibition of Chemical Weapons (OPCW) reached an agreement whereby an exception would be made for the part of the Rabta Industrial Facility that was never intended for chemical weapons production. That portion of the facility will be allowed to be used for civilian pharmaceutical production. See OPCW. 2004. *OPCW Executive Council Approves Recommendation to Allow for Conversion of Former Chemical Weapon Facility in Libya.* October 18. Available as of March 2009 at http://www.opcw.org/news/news/article/opcw-executive-council-approves-recommendation-to-allow-for-conversion-of-former-chemical-weapon-fac/.

[5] Libya did not maintain a state-sponsored biological weapons program. See Cirincione et al. 2005. 324 pp and Mahley. 2004.

To succeed, it will need to be an *integrated, cooperative, collaborative*, global enterprise that is *responsive, flexible, adaptable*, and able to respond to the new security threats that it will need to counter.

CTR 2.0 is likely to be characterized by smaller projects that not only seek to reduce threats but also have the goal of helping others prepare to prevent or respond to new threats. The report of the Review Panel on Future Directions for Defense Threat Reduction Agency Missions and Capabilities to Combat Weapons of Mass Destruction, March 2008 (hereafter referred to as the Carter-Joseph Report) concluded that the DOD CTR program should expand its focus on counterproliferation activities, including threat awareness, equipment, and consequence management training and exercises, aimed at building national and regional capacities.[6] These should become major program themes in future cooperative threat reduction efforts. Similarly, given new WMD proliferation threats from terrorism, the DOD CTR Proliferation Prevention Initiative should undertake a larger role along with enhanced and integrated efforts to provide assistance with export controls, border security, shutting down trafficking routes, and stemming piracy, which all can contribute to controlling WMD proliferation in meaningful ways.

Such efforts contribute to building capacities that can enable states to more effectively engage in efforts such as the Proliferation Security Initiative, the Global Initiative to Combat Nuclear Terrorism, and the implementation of United Nations Security Council Resolution 1540 (UNSCR 1540). Each of these new areas, however, will require the establishment of new or improved relationships with allies or partner countries. Just as a beachhead was established in Moscow in the early 1990s by taking the time to identify shared priorities and to codevelop a strategic approach, many new beachheads will have to be established in the future to initiate the next generation of cooperative threat reduction programs: CTR 2.0.

Finding 2-2: CTR 2.0 efforts will likely be smaller and distributed across a larger number of countries carefully targeted on the sources of new threats rather than the large, physical infrastructure dismantlement or construction projects that were the hallmarks of the programs in the former Soviet Union.

Defense and Military Contacts (DMC) Program has the potential to be an even more important element of CTR 2.0. The DMC Program started under

---

[6] Ashton B. Carter, Robert G. Joseph, et al. 2008. *Review Panel on Future Directions for Defense Threat Reduction Agency Missions and Capabilities to Combat Weapons of Mass Destruction*. Cambridge: Harvard University. Available as of March 2009 at http://belfercenter.ksg.harvard.edu/publication/18307/review_panel_on_future_directions_for_defense_threat_reduction_agency_missions_and_capabilities_to_combat_weapons_of_mass_destruction.html?breadcrumb=%2F.

CTR 1.0, but became disconnected over the years from the DOD CTR program goals it was initially designed to support. DMC activities build a partner's capacity and can be a mechanism for exploring and establishing relationships in new partner countries and regions. Senior officers at three of the Unified Combatant Commands support expanding the DMC Program and linking it strategically to other U.S. security assistance efforts. Current DMC Program examples include the following:

- Traveling Contact Teams (TCTs) for maritime interdiction and Nuclear, Biological, Chemical warning and detection
- Antiterror TCTs
- Military Police familiarization exchanges
- National Guard State Partnership Program familiarizations and contact visits
- Regional counterproliferation and counterterrorism exercises
- Disaster preparedness/consequence management TCTs
- Support for other regional security initiatives

There is also significant potential for building new relationships through existing civilian and military health and infectious disease programs, and through other scientific and technical collaborations that engage local expertise. This topic is explored in more detail in Chapter 4.

Finding 2-3: CTR 2.0 should include long-term relationship and capacity building that can be the basis for future cooperative threat reduction activities, through defense and military-to-military engagement and other peer-to-peer engagement, such as in science.

The new security environment requires that the USG reassess the eligibility criteria for USG CTR assistance. This should allow more effective leveraging of resources, help avoid duplication among agencies, and ensure that programs are prioritized as part of a national security strategy. Traditionally, factors that have been taken into account included the following:

- Is the security threat high and direct?
- Does assistance respond to one or more U.S. national security strategy priorities, such as compliance with treaties, fulfillment of existing nonproliferation or security agreements, or participation in new nonproliferation or security initiatives?
- Is the partner willing to cooperate and are they in critical need of technical or financial assistance?
- Is the partner willing to provide access to the key individuals, facilities, or materials?

- Are other countries or organizations willing to cooperate and provide additional technical or financial resources?
- Is there congressional authority and are there appropriations that will support the effort?
- Have NGOs or the private sector been engaged?
- Is there a probability of success?

The United States and others involved in CTR 2.0 should continue to prioritize programs according to these criteria, at the same time realizing that the priorities among criteria may change over time. It may not always be possible for CTR 2.0 to tackle the highest and most direct threat first, but it may be possible to make real progress on other important but slightly lower priority threats.

Finding 2-4: Traditional criteria for determining eligibility for cooperative threat reduction engagement may need to be adjusted to reflect the changing security environment.

There may be instances where there are open disputes with parts of a partner country government, but receptivity in other parts of the same government; there may be opportunities to engage where access to facilities is not as open as the United States or others would prefer, but where incremental forward progress and the development of trust demands flexibility. Programs under CTR 2.0 will challenge those who implement them to fashion new approaches to each set of circumstances, balancing the interests of all sides.

Finding 2-5: As the lessons learned from the Libyan experience make clear, to make cost-effective contributions to U.S. national security in the future, USG CTR programs must be less cumbersome and less bureaucratic in order to provide agile and timely contributions. They must take greater consideration of the needs and wants of reluctant partners even as we keep focused on core U.S. objectives.

DEVELOPING MEANINGFUL METRICS

> Most current metrics for tangible program success measure U.S. program performance, not the impact of the programs and measures of success in the cooperating countries, which arguably should be the more important focus.

CTR 2.0 will operate under congressional authorities and appropriations and must account for how it spends taxpayer dollars and demonstrate the national security benefit it produces. The metrics that have been used traditionally to produce the Nunn-Lugar Scorecard (see Appendix F) are not necessarily the right metrics for measuring the impact of CTR 2.0, for which intangible metrics—relationships and processes—will be harder to measure. Traditional, "hard" metrics are important program indicators, but do not necessarily capture some of the important high-value "soft" program results.

An essential issue in developing and using program metrics to understand and assess the success of USG CTR programs is to pair the specific metrics employed with the stated program goals and objectives. Each partner, however, may emphasize different goals and objectives even for the same project. It is essential that not only the goals and objectives be discussed among the partners during the development of programs, but also that the measures of program effectiveness be discussed and mutually agreed to *before* the initiation of the program. Some of these metrics may be oriented toward developing a sense of trust, facilitating a better understanding of threat perceptions and prioritization of risk, fostering sustained support for addressing threats, or engaging a more diverse group of experts to develop creative ways to address threats. In these cases, each partner may present a different set of indicators, which would independently evaluate the effectiveness of USG CTR programs from their own perspective. Discussing and comparing the results of program effectiveness against mutually agreed, but independently assessed, metrics may provide another valuable opportunity for strengthened engagement.

> Traditional "scorecard" metrics are too often very quantitative in nature and are not always adequate measures of a program's success.

Developing a new approach to metrics will also demand an appreciation by the Office of Management and Budget (OMB), department-based program auditors, and Congress that hard "scorecard" metrics, often very quantitative in nature, are not always adequate measures of a program's success. This is particularly true when relationship and capacity building are the objective. The current OMB Program Assessment Rating Tool,[7] which is used to measure program performance across the government, is particularly ill suited to this kind of evaluation.

There also has been a tendency in CTR 1.0 to define metrics from the U.S. perspective without incorporating metrics from the partner country's perspective. Since sustainability is an important element of decreasing threat and increasing security, meeting the partner's expectations will be another tool to

---

[7] The Program Assessment Rating Tool is used across the government to assess and improve program performance of federally funded programs. Its uniform design and approach to milestones makes it difficult to reflect progress in diplomatic negotiations and other areas where intangible results are important. See http://www.whitehouse.gov/omb/part_default/.

project whether a program or project has any likelihood of being sustained in the long term by the partner. Aspects of programs that can help measure the impact in the partner's environment could include the following:

- Contributions (in-kind and financial) of the partner country—Has the partner been engaged from the beginning stages in a way that gives them a sense of ownership and responsibility for the program? Is there a desire on the part of the partner to see the program succeed and to sustain the program into the future?
- Transparency—Has the partnership developed under the program enhanced levels of trust and helped both sides understand the other better, for example appreciating differing perspectives on threat and response?

There may be additional measures for specific programs such as for biological threat programs. Examples include:

- Measurable improvement in speed of response to outbreaks
- Improved quality of disease reporting
- More active regional engagement
- More scientific and technical collaborations of a strengthened nature

Similarly, rather than counting numbers of sensors installed or training sessions conducted, alternative metrics for nuclear security programs could include the following:

- Reduction in nuclear smuggling incidents
- Increase in the number of regular and realistic exercises of nuclear security response structures
- Reductions in nuclear materials stockpiles
- Consolidation of nuclear material storage sites
- Development of security cultures

Many of these kinds of metrics can only be measured over a period of years and will not satisfy the annual demands of the budget cycle, but there has to be recognition that forcing traditional metrics requirements onto programs designed to have long-term impact will not work. Different metrics need to be explored and adopted. The DOD CTR Biological Threat Reduction Program ran an exercise in the fall of 2008 to test the biological detection and surveillance system in Georgia that has resulted from a USG CTR program. The conclusions of that test should be reviewed as a possible model for future metrics design.

Metrics also need to be reviewed and updated regularly as situations and U.S. goals evolve. For example, a lesson learned from some of the DOE CTR programs was that as programs evolved to meet changing needs, metrics were not adjusted in parallel. Consequently, when the program achievements were reviewed, there was a mismatch between new program goals and old metrics, leaving room for misinterpretation and criticism.

One possible example of an analytical model to map or measure the relationship aspect of CTR 2.0 is Social Network Analysis (SNA).[8] SNA is a mathematical technique for analyzing sets of relationships between individuals and organizations. It has grown in sophistication and application over the past several decades and may be a useful tool for identifying tangible benefits of the relationships developed under cooperative threat reduction programs.

In addition, more attention needs to be paid to how metrics are linked to criteria for determining program priorities and program success. If flexible criteria are used to decide which program efforts go forward, then metrics have to be similarly flexible to reflect the key issues that were part of the policy decision. Calibrating metrics to criteria will require greater thought than has typically gone into developing USG CTR program metrics. For example, a modest program to redirect a group of scientists can have a major impact locally or even regionally, but the program probably should not be advertised as having broad global impact.

As a recent National Research Council report has concluded,[9] determining adequate measures of program effectiveness is particularly difficult when the goals and objectives of specific programs are largely unquantifiable, such as in relationship building and strengthening partnerships. As a means of attempting to address this challenge, many international organizations have sought to more fully employ *impact evaluations* as a means of understanding what has been most successful in their programs and why. While sophisticated social science methodology is often used to conduct impact evaluations,[10] the four essential factors in attempting to establish the success of a particular program are as follows: (1) clearly defining what is being evaluated; (2) identifying a set of desirable outcomes that might reasonably result from the program (based on realistic expectations as to what might be accomplished by specific programs); (3) determining specific indicators of success calibrated to the specific desired

---

[8] See, for example, Ulrike Gretzel. 2001. *Social Network Analysis: Introduction and Resources*. Available at http://lrs.ed.uiuc.edu/tse-portal/analysis/social-network-analysis/.

[9] National Research Council (NRC). 2008. *Improving Democracy Assistance: Building Knowledge Through Evaluations and Research*. Washington, D.C.: The National Academies Press.

[10] The NRC report notes that "impact evaluation is the term generally used for those evaluations that aim to establish, with maximum credibility, the effects of policy interventions relative to what would be observed in the absence of such interventions. These require . . . : collection of baseline data; collection of appropriate outcome data; and collection of the same data for comparable individuals, groups, or communities that, whether by assignment or for other reasons, did and did not receive the intervention." *Improving Democracy Assistance: Building Knowledge Through Evaluations and Research*. 47 pp.

outcomes; and (4) measuring program outcomes against these indicators. Employing relevant techniques of impact evaluations may better position USG CTR 2.0 to assess its successes, particularly when program goals and objectives are not easily quantifiable and do not translate easily onto a scorecard.

Finding 2-6: The traditional metrics of DOD (and USG) CTR success are often useful for program evaluation. Warheads or delivery systems and launchers destroyed, weapons materials secured, and contractor full-time equivalent on target are more concrete than just total dollars spent, but these metrics do not adequately reflect threat reduction impact or account for the value of potential CTR 2.0 engagement against new threats in this century. The challenge remains to find measureable performance indicators that capture the true value of important future successes that may be less tangible and more difficult to document. Efforts to contrive such measures, however, can result in burdensome and misleading data that may distort sound assessments of policy implementation. For example, the dollar value of locks and alarms procured, or even the number, is less important than the degree to which an institute plans, trains, and practices security against intruders and the "inside threat." These latter considerations are more important, but less transparent and measureable.

CTR IN THE 21st CENTURY: OBJECTIVES

What are WMD and other strategic threats to the United States?

> The universal availability of information and the highly dual-use nature of many technologies, particularly in the biological and chemical fields, make it possible to develop a WMD without having previously worked directly in a state WMD program.

National security threats during the Cold War were not confined to the bilateral superpower relationship, but they were defined largely by a known adversary and the threat of nuclear war and other WMD. Even though the relationship was far more complex than that, nostalgia exists for the days when the greatest threat could be defined by the Cold War paradigm of the United States versus the Soviet Union. Today, the United States faces security challenges from a wide range of countries with existing as well as latent WMD capability; from nonstate actors; and from countries known to be proliferators or where weak, unstable governments enable individuals or groups to proliferate with the tacit or active knowledge of the government. A strategy focused on immediate, direct threats may not be as wise or relevant in the 21st-century environment as understanding change and risk and how to keep risks from developing into threats.[11] This is the environment in which CTR 2.0 must operate. Some of the challenges to be considered include the following areas:

- **Nuclear Challenges** In addition to concerns about an expanding number of countries having declared nuclear weapon programs, there is increasing concern that the spread of nuclear technologies, particularly as a result of the increased global focus on nuclear power, will enable more countries to become latent nuclear states with the expertise, technology, and materials that can be rapidly converted to weapons use. As the use of nuclear power expands, cooperative government and industry efforts to develop more proliferation-resistant technologies, the possible establishment of an international fuel cycle program, and related efforts will all contribute to addressing these new challenges. Control over material is the final defensive line, but left as the only major defense, it can be weakened in the face of covert activity, third-party assistance, and treaty breakout. The IAEA, nuclear industry, cooperative research, and even military-to-military interactions can provide greater transparency and confidence building, but few insightful specific measures currently exist to support, monitor, or verify these capabilities. For example, military-to-military ties have weakened over the years in several countries of concern, sometimes replaced by NGO or Track II activities that may or may not compensate for a lack of peer-to-peer embedded engagement on a continuing basis. Even where military-to-military relations or other peer engagements are conducted, it is difficult to gauge how much they support USG CTR objectives, but they often do.
- **Chemical Weapons Challenges** In addition to countries with known or suspected chemical weapons programs, any country with a reasonably advanced chemical or petrochemical industrial sector has the latent capability to develop chemical weapons. Over the past several years, even what defines a "chemical weapon" has been challenged, for example, by the innovative use in Iraq by al Qaeda of chlorine gas tanks—either alone or in combination with improvised explosive devices (IEDs) to execute chemical attacks.[12] There has been international concern for many years over the lack of security at chemical facilities that could be targets for terrorist acquisition of components of a chemical weapon or simply releasing toxic industrial chemicals as a weapon. Because chemical weapons do not have the same potential strategic and catastrophic impact that nuclear and biological weapons do, there has been relatively little attention paid to developing a broad effort to increase security or to understand the

---

[11] See Appendix H for a comparison of the characteristics of Six Weapons Systems from the Perspective of a State of Terrorist Organization.

[12] Kirk Semple and Jon Elsen. 2007. Chlorine Attack in Iraq Kills 20. *New York Times*. April 16. Available as of March 2009 at http://www.nytimes.com/2007/04/06/world/middleeast/06cnd-iraq.html.

implications of advancing technology, which some experts told the committee is a serious oversight.

- **Biological Threats** The October 2001 anthrax attacks in the United States, the 2003 severe acute respiratory syndrome outbreak, and dramatic recent advances in biosciences and biodefense have spotlighted the destructive potential of biological agents. With strategic and catastrophic potential equal to nuclear weapons, the U.S. Congress has poured billions of dollars into efforts to defend against and respond to biological attacks. The issue has gained worldwide attention, and the threat of attack has served to drive the security and life science communities closer together, engaging experts in both human and agricultural health. USG CTR programs can enhance biosecurity by supporting a network of experts who are sensitive to international norms and national laws and who can bring their expertise even to troubled regions.
- **Dual-Use Expertise** State-sponsored WMD programs are no longer the only source of WMD threats. The universal availability of information and the highly dual-use nature of many technologies, particularly in the biological and chemical fields, make it possible to develop a WMD without having previously worked directly in a state WMD program. Educational opportunities abound, as demonstrated by the 2008 QS (Quacquarelli Symonds) list of top 100 universities.[13] Historically, most of the top-ranked universities have been in Australia, Canada, western Europe, Japan, New Zealand, and the United States; as the 2008 list shows, advances in education now make it possible to obtain a world-class education in life sciences and biomedicine, natural sciences, and technology at universities in China, India, Israel, Mexico, Russia, Singapore, South Korea, and Taiwan. These opportunities, combined with a wealth of training and other dual-use information available over the Internet, on CD, and through other channels, leave little chance for controlling the acquisition of WMD-related science, engineering, and technology.
- **Small Arms and Munitions Stockpiles** Combating the emergence of IEDs as a highly destructive terrorist weapon system is an urgent priority. According to the 2008 Small Arms Survey, "[d]iverted conventional ammunition, explosives, and military demolition items can be used in a wide range of IED types. . . . Large caliber ammunition, such as artillery shells and mortar bombs, are particularly useful for IED construction, because they contain relatively large quantities of explosive. In addition, military stockpiles frequently contain demolition stores, such as detonators, detonating cord, and plastic explosives that can greatly facilitate the construction of IEDs."[14] Although

---

[13] Top 100 Universities. 2008. Top 100 Universities. Available as of March 2009 at http://www.topuniversities.com/university_rankings/results/2008/overall_rankings/top_100_universities/.

[14] Adrian Wilkinson, James Bevan, and Ian Biddle. 2008. Conventional Ammunition in Surplus–A Reference Guide. *Small Arms Survey*. Geneva: Graduate Institute of International Studies. 17 pp. Available as of March 2009 at http://www.smallarmssurvey.org/files/sas/publications/b_series_pdf/CAiS/CAiS%20CH14%20IEDs.pdf.

IEDs are largely associated with Iraq, Afghanistan, and parts of the Middle East, the materials needed to feed the growing use of IEDs are available in many areas that are unstable, in the throes of civil or regional unrest, or are routes for trafficking. One estimate puts the amount of munitions in the Balkans alone at more than one million tons. In addition, the increasing availability and use of man-portable air defense systems presents another set of threats that must be combated.

- **Terrorism Threats**   The current U.S. list of Foreign Terrorist Organizations[15] consists of 44 groups, some with a national or regional focus and others that operate globally. Other lists, such as one published by the U.S. Committee for a Free Lebanon, name hundreds of organizations operating around the world, including in the United States. Terrorist organizations can access communications, transportation, and training, making it possible for today's local terrorist organization to reach far from its roots. Efforts are increasing to understand the roots and causes of terrorism, but much more needs to be understood. This report does not attempt to address the roots of terrorism, but there are cases where a CTR 2.0 relationship might help. In countries such as Pakistan, the lack of healthy military-to-military relations, once common decades ago, is being painfully felt now. Similarly, the lack of robust scientific and other peer-to-peer engagements has left a gap that terrorists have exploited successfully to recruit in several fields. This demonstrates widely how a strategy to oppose terrorism must look.

- **Border Security Threats and Trafficking (WMD, Drug, Contraband)**
There are growing concerns about evidence that trafficking routes, which traditionally move drugs, small arms, or people, can be utilized for the movement of WMD. The Department of Homeland Security's chief of intelligence analysis has been quoted as saying terrorists could use well-established smuggling routes and drug profits to bring people or even weapons of mass destruction to the United States. Drug lords in Latin America "already have well-established methods of smuggling, laundering money, obtaining false documents, providing safe havens and obtaining illicit weapons, all of which would be attractive to terrorists who are facing new pressures in the Middle East and elsewhere."[16] In a similar vein, a report released in November 2008[17] describes in detail how a key precursor for mustard gas was smuggled to both participants in the Iran-Iraq

---

[15] The State Department. 2008. Foreign Terrorist Organizations: Fact Sheet. Office of the Coordinator for Counterterrorism. Available as of March 2009 at http://www.state.gov/s/ct/rls/fs/08/103392.htm.

[16] Curt Anderson. 2008. U.S. Officials Fear Terrorist Links With Drug Lords: U.S. Officials Worry Islamic Extremists Could Form Alliances with Latin American Drug Lords. Miami: Associated Press. October 8. Available as of March 2009 at http://abcnews.go.com/US/wireStory?id=5986948.

[17] Jonathan B. Tucker. 2008. Trafficking Networks for Chemical Weapons Precursors: Lessons from the Iran-Iraq War of the 1980s. Occasional Paper No. 13 from the Center for Nonproliferation Studies. Monterey: Monterey Institute for International Studies. November.

war. Two independent trafficking networks purchased the precursor from a U.S. firm and shipped the material by circuitous routes to Iran and to Iraq. The report concludes that the same methods are being used today to smuggle precursors for nuclear, chemical, and biological weapons to rogue states and terrorist organizations. In the United States, oversight of exports is divided among the Departments of Commerce, Homeland Security, State, and Defense. Overseas, monitoring of trade in dual-use chemicals and biological materials (those that have legitimate commercial uses as well as utility in weapons production) is conducted by the Australia Group,[18] a group of nations that informally exchange intelligence information about potential violations of international regulations on trade in "scheduled chemicals" in the Chemical Weapons Convention (CWC). The process is reasonably effective, but is hampered by the nonparticipation of several emerging industrial powers. CTR 2.0 could play a role in enhancing the effectiveness of cooperative approaches to interdicting trafficking in weapons precursors of all sorts.

- **Failing and Failed States**   The 2008 economic crisis may have had its roots in the United States, but its global reach was swift and devastating in ways that have yet to be measured and with results that may present new threats to U.S. national security. Countries that already are struggling for control over territory, are dealing with internal conflict, or are coping inadequately with demands to provide for basic human needs may find themselves facing even greater challenges. The *Failed State Index*[19] presents a sober picture of potential future risk. In 2008, 177 countries were listed, 35 of which were considered to be "alert" states, including some that were mentioned to the committee by various experts as possible partner countries for future CTR 2.0 security engagement programs: Afghanistan, Democratic People's Republic of Korea, Democratic Republic of the Congo, Iraq, Lebanon, Pakistan, and Syria.
- **Cyber Threats**   News reports that hackers attacked the Web sites of the 2008 U.S. presidential candidates and broke into White House e-mail demonstrates again the vulnerability of the electronic universe. Some hackers may see penetrating government cybersecurity systems as the ultimate challenge, but it increasingly is a very serious business. Remotely disabling an opponent's radar system, as the Russians did in the August 2008 conflict with Georgia, can be an effective military tactic. On the other hand, generating false data that might indicate an attack is imminent when, in fact, that is not the case also can produce dangerous situations and a climate of tension and distrust.

The points above are not intended to be comprehensive, but illustrate the

---

[18] See http://www.australiagroup.net/en/index.html for additional information on the Australia Group.
[19] The Fund For Peace. 2008. *Failed State Index*. Available as of March 2009 at http://www.fundforpeace.org/web/index.php?option=com_content&task=view&id=99&Itemid=140.

orders of magnitude difference between managing the relatively stable and even predictable WMD threats of the Cold War, compared to the complex, multiple-risk contemporary environment. Even scientists working on the cutting edge of their disciplines cannot predict the future potential of rapidly evolving technologies, such as biotechnology and nanotechnology, let alone what happens when such technologies are combined. Experts also cannot predict where the next big technological innovation will appear or how rapidly it will evolve. The global reach of information now enables new discoveries to be assimilated and amplified rapidly. Instead of sharing a problem with one or two colleagues in a laboratory, scientists now can interact globally and instantaneously, accessing massively linked networks. Progress in such an environment will be exponential rather than linear, making predictability ever more difficult.

For governments to manage risk and respond to threats in this environment, they will have to work in a similarly more cohesive, integrated, networked fashion that operates in the same flexible, adaptive environment as the threats they are trying to counter. It will be increasingly difficult to solve challenges with the resources and capabilities of a single program, department, or even by the U.S. government acting unilaterally. CTR 2.0 cannot address the entire threat spectrum, but it is designed to reach beyond CTR 1.0's original focus.

Finding 2-7: The globalized environment will be characterized by the increasingly rapid spread of technology, major changes in how the traditional nation-state structure works or nongovernmental organizations engage, diffusion of threats, and changes in the nature of the threats (including the convergence of technology, new patterns of technologies, Internet-facilitated communications, and complex relationships).

The United States faces a broad array of WMD proliferation challenges around the globe. Reducing those threats in diverse political, economic, and technological environments will require a similarly diverse set of tools, approaches, and practices. The challenges—real and perceived—and the appropriate responses will be different from country to country and region to region. State-sponsored nuclear weapons programs are generally known or identifiable and can be addressed directly. The potential for loss of nuclear weapons or materials to rogue states or terrorist groups, although a more difficult challenge, can be reduced through security measures and multilateral cooperation. Chemical weapons programs and dual-use chemical industrial capabilities have physical signatures that may also allow identification, at a minimum, and possibly coordinated responses by the international community. Exploitation of toxic industrial chemicals and toxic industrial materials by terrorist groups or even lone actors requires a much different set of tools, best implemented by individual states (e.g., UNSCR 1540) or intelligence and security collaborations between allied states. Illicit biological weapons programs, once hidden in "closed cities" and possessing a relatively large footprint, are no longer necessary to develop a biological weapon. There may be no discernable signature provided by an offensive biological weapons program developing a highly contagious viral pathogen, for example.

> Each nation or region has a slightly different perception of the WMD proliferation threat; thus, there is a need for a graded degree of responses and broad potential for collaboration.

Just as challenges vary across weapons systems, they are also spread geographically. For example, a developed country might be the source of technologies or a rogue group of scientists, while a developing country may be the chosen location of a terrorist laboratory or simply a source for isolation of pathogens from nature. In response to a scenario like this, the former country may be willing to collaborate on preparation, response, and intelligence gathering, while the latter may seek capacity improvements to bolster its ability to identify outbreaks and secure its pathogen collections. Each nation or region has a slightly different perception of the WMD proliferation threat; thus, there is a need for a graded degree of responses and broad potential for collaboration.

Traditionally, the USG CTR programs have focused on WMD. Increasingly, the USG has recognized that radiological source security and munitions stockpile destruction and security are areas where some modest activity could have a significant impact on preventing an attack with a radiological dispersion device (or "dirty bomb") or an IED. It is sobering to think about what IED attacks in Iraq might be if DOD and DOE had not engaged early through a program to gather and secure radiological sources. Each of the weapons risk categories offers different opportunities to engage internationally, and all require a greater global awareness.

Efforts to address the nuclear nonproliferation challenge relies on various transparency tools, technical engagements, the work of the IAEA to implement the Nuclear Nonproliferation Treaty, and diplomatic interventions. Chemical risk reduction can be accomplished through intelligence on terrorist intent, security of industrial chemicals, activities of the Organization for the Prohibition of Chemical Weapons to implement the provisions of the CWC, and international collaboration in specific technical areas. Biological nonproliferation in the 21st century is unique in that international collaboration across the biorisk spectrum of infectious disease can focus our efforts where the capabilities to do harm with biology are the greatest and the potential for "intent" to harm is the most likely. At the same time, broad public health engagement and capacity building in the biological risk space can pay dividends by winning "hearts and minds" in communities where terrorists need the support of the local community to succeed. Appendix H provides a matrix for considering six areas of potential threat and how each might be viewed by a potential adversary.

Finding 2-8: Proliferation challenges and opportunities to control proliferation vary greatly geopolitically and across the three major WMD systems of concern as well as other areas of concern, and a diverse set of tools and approaches is needed to respond. Addressing these challenges will require the involvement of partners beyond the traditional government players and geostrategic allies.

CTR IN THE 21st CENTURY: OPPORTUNITIES

Partnership—The Critical Element

A key to future security will be the ability to engage new partners globally and to build a broad network committed to enhancing global security through engagement. This network of partnerships, in turn, can be a trip wire to warn us of potential dangers. Much of the success of this will depend on our ability to develop sustained, trusting human relationships. Although the committee can appreciate the desire for rapid, surgical, quick-impact actions, it does not believe that this kind of approach reflects the reality of the complex global environment that continues to evolve around us. Clearly, the United States and others should be prepared to respond rapidly to events and opportunities, but investing in relationships for global security must be viewed as a long-term commitment and one in which the *process of engagement* may sometimes be more important and yield more tangible results than the actual project.

Finding 2-9: Relationships on multiple levels with allies, threat-reduction partners, academe, NGOs, and others are necessary for effective engagement with countries and regions on nonproliferation activities.

Recent events have demonstrated the power of partnerships. In the aftermath of the August 2008 military action between Russia and Georgia, most high-level U.S.-Russian dialogue ceased. The U.S. nuclear cooperation agreement with Russia (also known as a "123 Agreement") was withdrawn from congressional consideration, and other important bilateral security topics were swept from the table by both countries. In the midst of this very public disagreement, however, USG CTR activities were allowed by both countries to continue. In contrast, following the September 2008 U.S. announcement of arms sales to Taiwan, China made a broad decision to curtail all security-related activities with U.S. government officials, canceling all scheduled bilateral events. The difference is striking, and although U.S. relations with Russia and China will recover over time, there is not yet as much substance in U.S. relations beneath the surface with China as exists with Russia; the habit of cooperation is not yet established. USG CTR programs play a limited but important role in building the foundations of long-term cooperation.

As CTR 2.0 is implemented, there also must be sensitivity to a response by some potential partners to working with the U.S. military when assigning roles under a strategic framework. Although the DOD CTR program has been successful in many environments, it is viewed with suspicion and distrust in others. DOD's attempts to engage directly with the Russian Ministry of Health, for example, were not successful, whereas it was able to engage successfully beyond the counterpart ministries of defense in Georgia and Uzbekistan. This unease can extend as well to NGOs that work as contractors for U.S. government programs. One NGO official commented to the committee that his organization prefers to receive funding for nonproliferation activities from the State Department rather than DOD because the source of funding can have a profound affect on how well the organization is received in a country where military affiliations and motives are distrusted. Russia and other countries have also expressed concerns about the use of DOD programs and personnel for intelligence purposes, a concern likely to be echoed by some new partners.

Finding 2-10: There are many potential partners and resources that can be employed for CTR 2.0 that currently are not being tapped.

In the CTR 2.0 model, partnership is a defining concept. In many respects, it will be even easier to establish this baseline with new partners than to manage the transition from assistance to a new cooperative framework with established partners, such as Russia. Even in the Russian case, committee members believe that the transition can be accomplished if it focuses on shared areas of security values and goals.

Finding 2-11: There is national security benefit in sustained partnerships, collaborations, and joint activities that link individuals and institutions in productive, mutually beneficial pursuits that can withstand political, economic, and other disruptions. This sustained engagement is the foundation for a "habit of cooperation."

New Partners to Support New Projects

The transition from CTR 1.0 to 2.0 will involve many changes in the operational patterns of DOD and other USG agencies and participants. Some of these changes, such as engagement with new partners, are inevitable and have already arisen in existing programs to varying degrees. The committee sees the potential for a more regular engagement of a broad range of nongovernment partners in CTR 2.0. There have been some successful DOD CTR examples of this already.

The committee believes greater collaboration and partnership between government (and international) security engagement programs and the private or nongovernmental sector, defined broadly as including industry, academe, and other organizations, could be particularly useful in program initiation and implementation. Organizations already operating in potential partner countries can also provide valuable insights into everything from local political dynamics to practical aspects of program implementation. In looking at these new partners, it is important to recognize that one group may be independent and function as full partners,[20] a second group may carry out designated tasks for a government program as subcontractors, and a third group may be a hybrid.

The academic community also may have much to offer. Since USG CTR began, many universities, academic institutions, and individual professors from many fields have had some role or contact with cooperative threat reduction programs. Their knowledge can be invaluable and their ability to engage on the ground may be indispensable. For example, a current project in Iraq involving Texas Tech University began under the auspices of a State Department nonproliferation program and subsequently received additional support from the university itself.[21] The two senior members of the Texas Tech faculty involved in the project have continued to provide their expertise to similar efforts elsewhere and are providing both an essential link to the scientific community and continuity for the project. In addition, many U.S. universities could partner with foreign universities. They may be able to provide unique insights into the local situation, as well as be potential cosponsors of meetings, workshops, or training programs.

The committee believes that industry has the potential to make a far greater contribution than it currently does to global security engagement. Early efforts to engage industry in DOD CTR were centered primarily on the Defense Conversion Program, which aimed to convert former Soviet military infrastructure to peaceful, civilian commercial production. This report will not cover the reasons why the Defense Conversion Program was eventually eliminated, but some of the program's efforts were successful and may be worth reviewing in the future. Industry should be tapped as a resource and partner as CTR 2.0 addresses threats that are inherently dual-use, particularly in the chemical industry and biotechnology sectors. Industry can also play a role as a barometer of local conditions that may have an impact on any programs in new partner

---

[20] For a description of activities undertaken by the Nuclear Threat Initiative, many of which have been in partnership with governments or international organizations, see www.nti.org.

[21] Richard Stone. 2008. Nuclear Control: Iraq Embarks on Demolition of Saddam-Era Nuclear Labs. *Science*. 321:5886. 188 pp. Available as of March 2009 at http://www.sciencemag.org/cgi/content/full/sci;321/5886/188?maxtoshow=&HITS=10&hits=10&RESULTFORMAT=&fulltext=carl+phillips+iraq+texas+tech&searchid=1&FIRSTINDEX=0&resourcetype=HWCIT.

countries. For example, 112 American Chambers of Commerce currently operate in 99 countries,[22] and generally have close ties both with the host country as well as with the U.S. embassy, and chambers of commerce representing the interests of other governments.

Another way to reach out to the U.S. private sector may be through the Overseas Security Advisory Council (OSAC), which held its 23rd Annual Briefing with former Secretary of State Condoleezza Rice on November 19, 2008. OSAC was started in 1985, as a forum to share ideas between the public and private sectors in recognition of the growing threat posed by international terrorism to U.S. citizens living and working abroad. The council, which now comprises a dozen federal agencies and more than 5,600 private-sector groups, meets annually to share ideas, and plays "an important role in helping to shape the world's view of America, how we maintain our security and how we engage with our neighbors in their countries."[23] Secretary Rice cited several areas where OSAC had played a role in 2008, including at the time of the bombing of the Marriott Hotel in Islamabad: "OSAC quickly gathered security information and shared that information, and that was used, in turn, to brief senior [State] Department officials and . . . chief security officers."[24] OSAC could play a role in the CTR 2.0 model.

Clearly, other governments and international organizations have played a strong role in CTR 1.0 and should continue to do so in CTR 2.0. The conceptual difference, however, is that in the CTR 2.0 model, new program engagements would begin by sharing information, identifying areas of risk and opportunity, and jointly planning responses. In the past, the United States has often been very parochial in its approach to program planning, and on many occasions this has had very negative consequences. An early example of this was the decision in 1992 to respond to Ukraine's objection to participating in the International Science and Technology Center (ISTC) WMD scientist redirection program based in Moscow. Because there were compelling reasons to keep Ukraine involved in broad CTR efforts, the United States agreed, but did not consult, with its ISTC partners before announcing its commitment to a new center—the Science and Technology Center in Ukraine. Because it had not been involved in the decision to pursue a separate center in Ukraine, the European Union (EU) opted not to participate and Japan made a similar decision. Although the EU eventually joined the Science and Technology Center of Ukraine several years after its creation, the situation might have been avoided or

---

[22] U.S. Chamber of Commerce. 2009. American Chambers of Commerce Abroad. Available as of March 2009 at http://www.uschamber.com/international/directory/default.

[23] Condoleezza Rice. 2008. Secretary's Remarks: Keynote Address at the Overseas Security Advisory Council 23rd Annual Briefing, November 19. Available as of March 2009 at http://vienna.osac.gov/page.cfm?pageID=4942.

[24] Ibid.

other solutions to Ukraine's objections could have been explored if the matter had been handled differently.

If the CTR 2.0 model is to work well, countries other than the United States will also need to adopt a commitment to consultation and coordination–a habit or culture of partnership. This will require diplomatic effort, but the existing bilateral, multilateral, and international channels of communication should provide ample opportunity for this.

CTR IN THE 21st CENTURY: LESSONS

The Shchuch'ye Example

The development, design, and construction of a chemical weapons destruction facility at Shchuch'ye in western Siberia is a good example of what is described above. In its 14-year history, the project has taken on many CTR 2.0 characteristics. It began as a typical CTR 1.0 project based on a bilateral agreement, and implementation was overseen by a U.S. prime contractor. However, when faced with obstacles, such as a congressional restriction that U.S. funds were to be used only for construction of the demilitarization facility and not for the essential supporting infrastructure, it became critical for DOD CTR to identify and engage new partners. The EU and 12 nations contributed funds, materials, and expertise for infrastructure construction at the facility (Box 1.1 in Chapter 1). Canada was particularly interested in the railroad construction project. It matched a $1 million challenge grant from the NTI with approximately $35 million to fund the construction of a rail line from the weapons storage area to the destruction facility.[25] Another important NGO partnership developed when Russian public opposition to the project developed over environmental concerns. Global Green USA, the U.S. affiliate of the Russian NGO, Green Cross, was engaged to survey public opinion and establish a public involvement program.[26] Global Green has now opened and managed more than a dozen local outreach offices at chemical (and nuclear) CTR sites to facilitate threat reduction projects. This was done in coordination with Green Cross Russia and Green Cross Switzerland, and with the support of about a dozen Global Partnership members.[27]

---

[25] Laura Holgate. 2008. Discussion at CTR Committee Meeting #2, July 8.

[26] Igor Khripunov and G. W. Parshall. Nongovernmental Actors in U.S. and Russian Chemical Demilitarization Efforts. *Demokratizatsiya*. 9:10. 422-457 pp.

[27] More information on the Shchuch'ye effort can be found at Global Green U.S. 2009. Security and Sustainability. Available as of March 2009 at www.globalgreen.org/wmd.

The Vinca Example

Another example is the role played by NTI in removing nuclear materials from the reactor in Vinca, Serbia. When Serbian president Slobodan Milosevic was ousted, the U.S. government decided to include Project Vinca in its package of support, but would only deal with the highly enriched uranium (HEU) at the reactor, which it viewed as the immediate proliferation threat. NTI, however, saw a link between the 40 kilograms of HEU and the opportunity to remove spent fuel from the unstable environment. A seven-party action plan was devised involving the IAEA, the Department of State, DOE, the Serbian government, a laboratory, NTI, and the U.S. Embassy in Belgrade. Because NTI was not covered for liability under the Memorandum of Understanding that governed the project, the work was managed directly by the IAEA. NTI provided funding to the IAEA, becoming the first nonstate donor to the IAEA.[28] The project helped highlight the need to phase out the use of HEU in civilian facilities.

This joint and highly collaborative project had several successes by the summer of 2008: the removal of fresh fuel in 2002 through a transparent, seven-party process and removal of nearly all the spent fuel. In addition, the IAEA raised more funding, and the project helped highlight the need to phase out the use of HEU in civilian facilities.

What Makes NGOs Different?

NGOs typically do not depend on government money and NGO officers are not government officials. This leaves NGOs able to criticize government approaches, be more responsive to recipient concerns, act quickly, remain flexible, and be more willing to accept risks. NGOs are also unfettered by federal acquisition regulations and can design any kind of contracting mechanisms that will meet requirements. An NGO can pick and choose among different projects and focus on those it feels are closest to the organization's mandate, and can bring in expertise either through their staff or through networks of experts. In many cases, NGO boards include international experts that lend a high level of prestige and credibility that helps gain access.

There can be possible disadvantages to working with NGOs as well. For example, there is the risk that a foreign government might think that an NGO is speaking on behalf of the U.S. government. NGOs may not have the same level of accountability as the government, may not have the necessary technical expertise, or may even work at cross purposes with the government.

---

[28] Nuclear Threat Initiative. Press Release. NTI Commits $5 Million to Help Secure Vulnerable Nuclear Weapons Material. Nuclear Threat Initiative, August 23, 2002. Available as of March 2009 at http://www.nti.org/c_press/release_082302.pdf.

NGOs Then and Now

The potential for developing constructive and effective working relationships with NGOs is higher now than it was in the early days of CTR, when concepts were new and every program was a learning process. Now, several NGOs have boards and staffs that were intimately involved with USG CTR efforts and understand the underlying security issues, congressional dynamics, and budgetary constraints.[29] Not only do the individuals involved have requisite expertise, their organizations have also gained reputations in the field for carrying out programs and activities in support of the CTR mission.

NGOs, especially those with a demonstrated track record, can work with USG CTR programs in different ways that need to be explored more systematically.

- Pioneers, or "wedge strategies" – NGOs can take risks that governments may not be ready to take. This may be particularly important when exploring the potential to engage a new partner country.
- Analysis leading to catalysis – NGOs can undertake analytical efforts designed to solve a problem, including undertaking independent approaches to resolve political questions that are impeding action.
- Design sharing – NGOs can work with the government on projects where each has a defined role.
- Gap filling – NGOs can fill gaps when the government either does not see the gap or cannot participate.
- Deal closer – NGOs can pressure groups to get them to honor their agreements.
- Setting benchmarks and evaluating – NGOs can be an independent voice to evaluate programs.
- Safe convener – NGOs can also get agencies or even countries around a table when there are disagreements or frictions that prevent constructive dialogue.

Programs that address threats to national security will predictably involve types of information that may not be appropriate to share with nongovernmental organizations. Judgments made based on sensitive diplomatic, security, proprietary, or privacy issues may need to guide aspects of program design, implementation, and oversight, and it may not be possible to be completely open and transparent. However, a great deal can be accomplished if guidelines set appropriate boundaries. There is already a substantial history of successful nongovernmental partnerships with organizations such as the National Acad-

---

[29] For example, the Civilian Research and Development Foundation (www.crdf.org) and the Nuclear Threat Initiative (www.nti.org) both have board and staff members who have long been active in CTR activities.

emies, the Nuclear Threat Initiative, the Center for Strategic and International Studies, the Henry L. Stimson Center, the Global Security Partnership, and others.

Sustainability

The issues of financial and other commitments from CTR program partners link directly to both integrating partners into initial program development and designing programs to be sustainable. If the United States expects increased financial commitments from partner countries—both those in which programs will be implemented and those who can contribute to implementation—it cannot maintain the paternalistic approach that has characterized past programs. The CTR 2.0 goal of building partnerships through shared program development, mutually agreed-upon goals, and joint funding—defined to include in-kind as well as direct contributions—has a much greater likelihood of leading to sustainable, mutually beneficial activities.

In the case of Russia, where the committee believes cooperative work should continue, the issue of funding could be addressed in three ways. First, no systematic mechanism was ever developed that adequately reflects the contributions that Russia (and other former Soviet states engaged in USG CTR activities) have made. Although these contributions were largely in-kind, they were often significant. Some efforts were made through the ISTC's scientist redirection program, for example, to require participating institutes to cover part of every project's costs. Overhead for projects was capped at a very low percentage and only a portion of nontechnical staff involved in any project could be covered, leaving it to the institute—or the government—to cover the remaining overhead and personnel costs. These and other in-kind contributions need to be captured and acknowledged to encourage future partnerships, particularly in environments where cash contributions are unlikely.

Second, although a country may need extensive support at the beginning of a project, economic factors may change that would allow that country to take on a higher share of the implementation burden. This is now the case with Russia, but no provision was ever included in agreements that would allow for this. The ability to assume a higher burden for in-country costs, however, should not signal the end of the partnership. The Libya example shows that having the financial resources to support a project does not mean that a partner does not need technical or other assistance, or should not continue to be engaged as part of a sustainability strategy.

Third, a partner country that initially receives assistance can evolve into a future implementation partner. This may be particularly useful where it may be difficult for the United States to engage a new partner. Chemical weapons destruction in the Middle East is a possible example. If the United States works with Iraq to destroy its remaining chemical weapons stockpile and trains an

Iraqi chemical weapons disposal unit, it may be possible to develop a project where the Iraqi unit—with continued U.S. technical support—engages with other countries in the Middle East on chemical weapons destruction issues rather than the United States. Along similar lines, it is hard to imagine any kind of security engagement program in North Korea that does not involve Russia. When new efforts are being designed, a useful planning exercise would be to think forward about the role that a new engagement partner might play in the future.

The committee also believes that fundamental changes in circumstances, such as those that have taken place with Russia, need to be governed by frameworks that can reflect a rebalancing of the relationship. Whether or not this is under a nuclear cooperation agreement or some other framework matters less than having a legal structure that covers not only what the United States and Russia do bilaterally, but also what they might do together elsewhere.

The committee also observed that DOD and DOE have taken steps to factor sustainability more specifically into program implementation. A February 2007 Government Accountability Office report states that "during our visit to Russia, officials at three of the four civilian nuclear research institutes we visited told us that they are concerned about their sites' financial ability to maintain U.S.-funded security upgrades after . . . DOE financial support ends in 2013."[30] DOD has announced plans to halt funding for analogous activities in 2011. In April 2007, the DOE's National Nuclear Security Administration announced that it had reached a nonbinding agreement with Russia's Federal Atomic Energy Agency (Rosatom) on a plan for Russia to sustain U.S.-funded security upgrades at nuclear material sites after DOE ceases its financial support.[31] Separate discussions have reportedly taken place with regard to Russia sustaining U.S.-funded work performed at sites with nuclear warheads and at nuclear material sites controlled by other agencies. Additional, legally binding bilateral arrangements may have a useful role to play in ensuring that operation and maintenance of U.S.-funded security upgrades continue to receive the requisite levels of Russian funding after 2011 and 2013. Consideration could also be given to how to address similar sustainability challenges with respect to U.S. and other international efforts to engage former Soviet WMD scientists in

---

[30] Government Accountability Office. 2007. *Progress Made in Improving Security at Russian Nuclear Sites, but the Long-term Sustainability of U.S.-Funded Security Upgrades Is Uncertain*. GAO-07-404. 26-27 pp. In section 3156(b)1 of the National Defense Authorization Act of 2003, Congress directed as follows: "The Secretary of Energy shall work cooperatively with the Russian Federation to develop, as soon as practicable but no later than January 1, 2013, a sustainable nuclear materials protection, control, and accounting system for the nuclear materials of the Russian Federation that is supported solely by the Russian Federation."

[31] National Nuclear Security Administration (NNSA). 2007. *U.S. and Russia Agree to Sustain Security Upgrades at Nuclear Material Facilities: Agreement Helps to Ensure that U.S. Investments Will be Maintained*. March 29. Available as of March 2009 at http://nnsa.energy.gov/news/1131.htm.

non-WMD-related employment, and CTR 2.0 activities both within and outside the former Soviet Union.

One of the lessons learned from USG CTR programs operating in the former Soviet Union is that sustainability has to be part of the original program plan, not something that is an afterthought. Under CTR 2.0, advance planning carried out in cooperation with the partner should be standard practice. Other agencies and entities that provide foreign assistance—such as the U.S. Agency for International Development, the Millennium Challenge Corporation, and other nongovernmental organizations—have developed their own mechanisms for promoting sustainability after the conclusion of donor funding. Reaching out to such foreign assistance providers in order to identify and, where appropriate, replicate their best practices would contribute to DOD and other USG CTR program effectiveness and may offer new opportunities for program leveraging.

Finding 2-12: Engagement programs are more effective and have a higher likelihood of being sustained if they are developed in partnership with the engaged country, are tailored to the region, and are seen as beneficial to both partners.

Recommendation 2-1: The White House, working across the executive branch and with Congress, should engage a broader range of partners in a variety of roles to enable CTR 2.0 to enhance global security. At a minimum this will require

- *Becoming more agile, flexible, and responsive*
- **Cultivating** *additional domestic and global partners* **to help meet our goals**
- **Building mutually beneficial** *relationships* **that foster** *sustained cooperation*

EVOLVING FROM CTR 1.0 TO CTR 2.0

Implementing CTR 2.0 will be an incremental process. The United States needs to continue to address old challenges as it organizes to meet new ones. A well-crafted plan can ensure that good practices of CTR 1.0 are embraced and carried forward, and unproductive ones are shed. One possible model is to build on the current National Security Council–Homeland Security Council bioengagement coordination efforts. Although relatively new, it already has learned lessons about the challenges of getting diverse communities to work together productively and is a first step toward CTR 2.0 implementation. Other lessons learned should be incorporated into the transition process as well. The committee was told that DOE's National Nuclear Security Administration has developed a useful methodology for prioritizing countries that may participate in the Megaports Program.[32] This methodology might be applied across any number of programs or modified as appropriate. Another promising example is a new joint project between the U.S. Department of Agriculture, DOD CTR, and a host of other partners. The program design and funding engages Canada, the European Union, the Russian Federation, the United Kingdom, and the United States as partners to establish an animal health disease surveillance system in the Stavropol region of Russia. The project aims to coordinate all efforts from the beginning of engagement to ensure an efficient use of funds and maximum participation from a leading group of international scientists and veterinarians. In addition to supporting a veterinary disease monitoring system in an important geographic nexus between Russia and Europe, the project also aims to identify multiple subsequent opportunities for Russian Federation veterinary scientists to engage in long-term collaborative research projects with equivalent scientists based in Canada, the European Union, the United Kingdom, and the United States. Parts of this project were built on program concepts developed by the Nuclear Threat Initiative.

Finding 2-13: There needs to be a distinct transition plan to move between the current cooperative threat reduction programs and CTR 2.0.

SUMMARY OF CHAPTER FINDINGS AND RECOMMENDATIONS

Finding 2-1: CTR 1.0 was a highly creative response to unique security challenges and geopolitical changes in the former Soviet Union. The new threats we face require similar innovation to create CTR 2.0. Coordination and leadership from the White House will be required, and relevant departments and agencies will need to engage to ensure that there is a clear connection between the policy intent and program implementation, as in the case of Libya. To succeed, it will need to be an *integrated*, *cooperative*, *collaborative*, global enterprise that is *responsive*, *flexible*, *adaptable*, and able to respond to the new security threats that it will need to counter.

Finding 2-2: CTR 2.0 efforts will likely be smaller and distributed across a larger number of countries carefully targeted on the sources of new threats rather than the large, physical infrastructure dismantlement or construction projects that were the hallmarks of the programs in the former Soviet Union.

Finding 2-3: CTR 2.0 should include long-term relationship and capacity building that can be the basis for future cooperative threat reduction activities,

---

[32] National Nuclear Security Administration. Megaports Initiative. Department of Energy. Available as of March 2009 at http://nnsa.energy.gov/nuclear_nonproliferation/1641.htm.

through defense and military-to-military engagement and other peer-to-peer engagement, such as in science.

Finding 2-4: Traditional criteria for determining eligibility for cooperative threat reduction engagement may need to be adjusted to reflect the changing security environment.

Finding 2-5: As the lessons learned from the Libyan experience make clear, to make cost-effective contributions to U.S. national security in the future, USG CTR programs must be less cumbersome and less bureaucratic in order to provide agile and timely contributions. They must take greater consideration of the needs and wants of reluctant partners, even as we keep focused on core U.S. objectives.

Finding 2-6: The traditional metrics of DOD (and USG) CTR success are often useful for program evaluation. Warheads or delivery systems and launchers destroyed, weapons materials secured, and contractor full-time equivalent on target are more concrete than just total dollars spent, but these metrics do not adequately reflect threat reduction impact or account for the value of potential CTR 2.0 engagement against new threats in this century. The challenge remains to find measureable performance indicators that capture the true value of important future successes that may be less tangible and more difficult to document. Efforts to contrive such measures, however, can result in burdensome and misleading data that may distort sound assessments of policy implementation. For example, the dollar value of locks and alarms procured, or even the number, is less important than the degree to which an institute plans, trains, and practices security against intruders and the "inside threat." These latter considerations are more important, but less transparent and measureable.

Finding 2-7: The globalized environment will be characterized by the increasingly rapid spread of technology, major changes in how the traditional nation-state structure works or nongovernmental organizations engage, diffusion of threats, and changes in the nature of the threats (including the convergence of technology, new patterns of technologies, Internet-facilitated communications, and complex relationships).

Finding 2-8: Proliferation challenges and opportunities to control proliferation vary greatly geopolitically and across the three major WMD systems of concern as well as other areas of concern, and a diverse set of tools and approaches is needed to respond. Addressing these challenges will require the involvement of partners beyond the traditional government players and geostrategic allies.

Finding 2-9: Relationships on multiple levels with allies, threat-reduction partners, academe, NGOs, and others are necessary for effective engagement with countries and regions on nonproliferation activities.

Finding 2-10: There are many potential partners and resources that can be employed for CTR 2.0 that currently are not being tapped.

Finding 2-11: There is national security benefit in sustained partnerships, collaborations, and joint activities that link individuals and institutions in productive, mutually beneficial pursuits that can withstand political, economic, and other disruptions. This sustained engagement is the foundation for a "habit of cooperation."

Finding 2-12: Engagement programs are more effective and have a higher likelihood of being sustained if they are developed in partnership with the engaged country, are tailored to the region, and are seen as beneficial to both partners.

Recommendation 2-1: The White House, working across the executive branch and with Congress, should engage a broader range of partners in a variety of roles to enable CTR 2.0 to enhance global security. At a minimum this will require

- *Becoming more agile, flexible, and responsive*
- Cultivating *additional domestic and global partners* to help meet our goals
- Building mutually beneficial *relationships* that foster *sustained cooperation*

Finding 2-13: There needs to be a distinct transition plan to move between the current cooperative threat reduction programs and CTR 2.0.

3

The Form and Function of Cooperative Threat Reduction 2.0: Engaging Partners to Enhance Global Security

The committee identified several key Cooperative Threat Reduction (CTR) 2.0 elements and examples of new engagement opportunities. These examples are not intended to be comprehensive or prescriptive, but are meant to stimulate thinking about how CTR 2.0 can be implemented.

KEY ELEMENTS OF CTR 2.0

Global Security Engagement   CTR 2.0's objective is to enhance global as well as U.S. national security, recognizing that reducing threats to other nations has direct benefits to U.S. security. Global security engagement assumes that new partners participate not only because they confront or represent some level of threat, but also because they are security partners. This partnership will be reflected in longer-term efforts to build relationships and capacity.

The committee recognizes that not all new partners will be fully engaged at the outset or will even be fully cooperative. Similarly, partner attitudes may shift over time for political or other reasons, as has been the case with Russia. The challenges being addressed, however, may be so compelling that the engagement should proceed, even if partner cooperation is not as complete as might be desired.

Clear Strategic Plan   To advance substantially from what currently exists, CTR 2.0 must have a clear strategic plan and strong senior leadership. These are core requirements.

Cooperative Network   A set of programs and projects will be implemented by the United States in cooperation with a network of countries, international organizations, and nongovernmental partners. The goal is to prevent, reduce, mitigate, or eliminate common contemporary threats to security and prepare for future threats. The United States has such a wide range of assets that can be applied to CTR 2.0 that effective implementation will require *strong, high-level, central leadership*.

Partnership as the Basic Mechanism for Cooperation   Partnership in CTR 2.0 will mean that the countries participating must be ready to discuss and potentially support a rational division of responsibility for

- *Project leadership*, including project definition and planning
- *Management*, including project organization, implementation, and oversight
- *Resources*, including personnel, technical capability, financial, and in-kind contributions

A Creative, Flexible Approach to the Form and Substance of New Engagements   A creative and flexible approach will be needed both to developing the form and to developing the substance of engagements, as well as to the metrics used to measure these.

- **Form**   CTR 2.0 will be capable of both long-term programmatic engagements and rapid response. Although both are possible under CTR 1.0, the committee believes that there should be more flexibility in programs across the U.S. government. Piggybacking or comingling funds, allocation of funding across U.S. government programs, the flexibility of funds, new approaches to contracting, and other issues are dealt with in more detail in Chapter 4.
- **Substance**   CTR 2.0 will look broadly at how it can support both traditional cooperative threat reduction missions focused on weapons of mass destruction (WMD) as well as new threats such as countering WMD terrorism and similar challenges. In this context, building capacity may be an important component, both in global commitment to security and in the ability to detect and respond to events.

Various programs under CTR 1.0 supported important arms control treaty implementation commitments. CTR 2.0 will continue to support these activities, but will also look specifically at ways to support new and expanded multilateral and international security instruments, such as the Group of Eight Global Partnership (G8 GP), the Proliferation Security Initiative, the United Nations Security Council Resolution (UNSCR) 1540, and the Global Initiative to Combat Nuclear Terrorism (GICNT). The use of CTR 2.0 could help engage other countries as more active and effective participants in this new generation of security efforts.

Coordinating CTR 2.0

CTR 2.0 requires a much higher degree of coordination than currently exists in the United States, or between the United States and other partners. Coordination is also one of the key points that Congress asked to have considered in this report. The importance of coordination was also noted by the United Nations when it passed UNSCR 1810[1]:

> Resolution 1810 (2008) encourages the 1540 Committee to work more closely, in its outreach activities, with global and regional intergovernmental organizations, and arrangements within and outside the United Nations system to foster the sharing of experience, create forums for discussion and develop innovative mechanisms to achieve the implementation of the resolution.[2]

The committee heard a consistent and strong emphasis from many U.S. and international experts on the need for a cohesive strategic approach as the Department of Defense (DOD) and other U.S. government (USG) CTR programs become global. One senior official of a G8 GP country commented that the United States tends to "move out when it sees opportunities and go it alone on a lot of issues. It [the United States] can do a lot, but it cannot do everything. We need to work out how to do things in a complementary way, *before* we begin approaching new countries."[3]

As CTR 2.0 programs are implemented, they will need to take into account the myriad of other programs, organizations, and conditions in new high-priority engagement areas. Regional development banks, and assistance programs from countries and organizations that were not part of the calculus in the former Soviet Union (FSU) may become new partners. Other competing national and regional priorities, such as basic health, water, and food needs may limit how much can be done and in what time frame. Each new effort must begin with a clear strategy that assigns specific roles to U.S. government departments and programs and identifies the appropriate resources and capabilities for the task. This will be even more important as programs are implemented in the face of a deepening global economic crisis in which security may take a backseat to providing a population with the basic necessities of life.

Policy makers also must have reliable data on existing programs to develop an effective strategy. In 1991, the authorizing legislation for the cooperative

---

[1] UNSCR 1810, adopted April 25, 2008, extends the 1540 Committee mandate for three more years and calls on the 1540 Committee to intensify its efforts to promote the full implementation of UNSCR 1540. Available as of March 2009 at http://www.securitycouncilreport.org/atf/cf/%7B65BFCF9B-6D27-4E9C-8CD3-CF6E4FF96FF9%7D/Terrorism%20SRES1810.pdf.

[2] U.N. Security Council 6015th Meeting. November 12, 2008. New York, S/PV.6015. 4 pp. Available as of March 2009 at: http://www.securitycouncilreport.org/atf/cf/%7B65BFCF9B-6D27-4E9C-8CD3-CF6E4FF96FF9%7D/Terrorism%20SPV%206015.pdf.

[3] Mary Alice Hayward. 2008. Discussion at Committee Meeting #1. May 21.

threat reduction and humanitarian assistance programs, the FREEDOM Support Act (Public Law 102-511), established a "Coordinator of U.S. Assistance to the Former Soviet Union." This provided a central point in the Department of State that coordinated and monitored humanitarian and security assistance budgets and program implementation across all agencies. The database that once existed in that office is no longer maintained regularly, making it very difficult to see where there are program overlaps or gaps, or where programs could be integrated. The G8 GP has tried to maintain a database and several nongovernmental organizations (NGOs) attempt to track budgets for CTR programs, but these do not compensate for the lack of a comprehensive U.S. government tracking system.

Other countries, international organizations, NGOs, the academic community, and industry will also have insights into issues that can materially affect the success of future security engagement efforts and can provide important program data. If marshaled effectively, these diverse resources can increase the probability of program success and sustainability. A high degree of leadership and coordination within the U.S. government, and from the U.S. government with partners inside and outside the United States, will be required. The committee has not seen evidence that a model currently exists for this level of cooperative and collaborative interaction.

Finding 3-1: The lack of a government-wide tracking program for USG CTR programs that cross agency budgets impedes the U.S. government's ability to develop a strategic approach to CTR 2.0.

NEW OPPORTUNITIES FOR ESTABLISHED PARTNERS

Changing political dynamics may also have a profound impact on where and how CTR 2.0 programs are conducted. The tensions that have developed between the United States and Russia since the August 2008 Russian conflict with Georgia are an example. Russia remains a major recipient of USG CTR support and is the primary beneficiary of programs under the G8 GP. In addition to being a beneficiary, Russia could integrate that experience into approaches for global security engagements in new regions. Long-term ties between Russia (and in some cases the FSU) and countries such as the Democratic People's Republic of Korea and Iran may make Russian participation indispensable if engagement opportunities open in those countries. Similarly, Russia's educational ties with countries in the former Soviet sphere of influence may provide unique links that could be important in future security engagement efforts.

The Russians should be able to bring important insights to CTR 2.0 that can help inform and shape future approaches. Temporary political perturbations should not be allowed to disrupt or curtail efforts to complete, continue, and initiate threat reduction programs in Russia or to seek Russia's participation in pursuing threat reduction in other countries. Successful CTR projects in Russia, such as the Russian Methodological and Training Center at Obninsk[4] and the Animal Breeding Facility at the Pushchino Research Center, might serve as models for global efforts.[5]

THE ROLE OF LEADERSHIP

In August 2007, President George W. Bush signed into law a bill implementing the recommendations of the 9/11 Commission.[6] The law provides for the creation of a special White House office headed by the United States Coordinator for the Prevention of Weapons of Mass Destruction Proliferation and Terrorism, indicating Congress's recognition of a need for greater leadership and integration of efforts across the U.S. government. The committee also recognizes that some form of strong central leadership will be essential to the successful implementation of CTR 2.0. We applaud a recent effort that we believe epitomizes the spirit of CTR 2.0. This interagency effort, "United States Bioengagement Strategy," led by the National Security Council–Homeland Security Council (NSC-HSC) and begun in 2008, is a possible model for USG CTR's evolution. It is different because it encompasses security and nonsecurity agencies and programs to explore how they all can contribute to a common strategy. Beginning with this foundation in biological engagement, the NSC-HSC team could reach out even more broadly to traditional and nontraditional partners, possibly focusing on one country as a pilot project. Once the system has been established and the mechanisms have been defined, other working groups could develop similar models, working with different challenges in different countries and regions, to create the network we call CTR 2.0.

Another possible approach is proposed by the Project on National Security Reform (PNSR),[7] which recommends fundamental reforms in the organization of the U.S. national security system similar to what the Goldwater-Nichols Act[8] did for the U.S. military in the 1980s. The project's proposals are based on case studies that "assess a series of events and developments that would shed light on

---

[4] A brochure describing the activities at the Russian Methodological and Training Center is available as of March 2009 at http://www.nti.org/db/nisprofs/russia/fulltext/rmtc/rmtc1.htm.

[5] A brochure describing the activities of the center is available as of March 2009 at www.fp7-bio.ru/konferencii/v-international-symposium/pushchino-scientific-centre/at_download/file.

[6] Implementing Recommendations of the 9/11 Commission Act of 2007, P.L. 110-53.

[7] Project on National Security Reform. 2008. Available as of March 2009 at http://www.pnsr.org/web/page/682/sectionid/579/pagelevel/2/interior.asp.

[8] U.S. Code: Title 10,111. Executive Department, Title 10 - Armed Forces/Subtitle A - General Military Law/Part I - Organization and General Military Powers/Chapter 2 - Department of Defense. Available as of March 2009 at http://www.law.cornell.edu/uscode/search/display.html?terms=goldwater&url=/uscode/html/uscode10/usc_sec_10_00000111----000-notes.html.

the past performance of the United States Government in mitigating, preparing for, responding to, and recovering from national security challenges."[9] Some of the questions that guided the case studies in that report reflect the fundamental issues identified by the committee in its study of CTR programs: Were U.S. government efforts integrated and guided by an overarching strategy or were they ad hoc; and how well did the agencies and departments work together? The PNSR released its findings and recommendations to the White House and to congressional leaders in November 2008, and Volume 1 of its case studies in September 2008. The PNSR report contains sweeping themes and recommendations. The committee identified those that are fully compatible with the CTR 2.0 concept. These include the following:

- Adopting new approaches emphasizing integrated effort, collaboration, agility, and a focus on national missions and outcomes. This point includes several recommendations including one that would prescribe in statute the national security roles of each department and agency, especially those that have previously been viewed as part of the national security system. This would solve a problem that is addressed later in this report.
- Establishing clear White House authority for national security strategy coordination across the government and providing the resources to carry out this function.
- Creating interagency teams to manage national security issues.
- Revising the budget process to better link resources to national security goals.
- Improving the ability to develop and share information across national security agencies.
- Building a partnership between the executive and legislative branches.

A similar approach with White House leadership and interagency collaboration was proposed in the 2007 report on the future of the Biological Threat Reduction Program of the Department of Defense.[10] Whatever approach is ultimately adopted, its goal should be to eliminate the overlap and duplication that exists in CTR 1.0. The committee was told by an officer in one of the Unified Combatant Commands about a set of visits during 2008 to a Central Asian country by two different programs; one providing border security assistance and the other providing counternarcotics trafficking assistance. Both programs

---

[9] Richard Weitz, ed. 2008. *Project on National Security Reform: Case Studies Volume 1.* Available as of March 2009 at http://www.pnsr.org/data/files/pnsr%20case%20studies%20vol.%201.pdf.
[10] National Research Council. *The Biological Threat Reduction Program of the Department of Defense: From Foreign Assistance to Sustainable Partnerships.* Washington, D.C.: The National Academies Press. 54 pp. Available at http://www.nap.edu/catalog.php?record_id=12005 as of March 2009.

were dealing with the same agencies in the partner government, but unfortunately neither knew about the other's efforts.

Even with strong leadership from the White House, no new effort will succeed without the active and committed support of cabinet secretaries and other senior officials from all relevant agencies. It is difficult, however, to sustain senior-level engagement over the longer term. One possible solution would be to have regular White House-led reviews, perhaps on a biannual schedule, to drive higher-level attention and coordination.

Finding 3-2: Responding to the new global security challenges requires a new model of interagency leadership. CTR 2.0 will function most effectively with strong leadership from the White House, and with the active involvement of relevant departments and agencies.

Recommendation 3-1: CTR 2.0 should be directed by the White House through a senior official at the National Security Council and be implemented by the Departments of Defense, State, Energy, Health and Human Services, and Agriculture, and other relevant cabinet secretaries.

HAVING THE RIGHT TOOLS

USG CTR currently has a substantial array of programs and resources, but new engagements may require new tools or old tools used in a new way. Although CTR 1.0 programs encountered problems implementing programs in the FSU because of difficult economic times or social and political stress, these may be minor compared to the challenges of engaging countries like Afghanistan or Pakistan. For example, where Russia and other countries of the FSU had well-educated populations and adapted quickly to the technology used in many USG CTR projects, it may be a challenge to find user-friendly and environmentally appropriate approaches for countries that are less developed. Officers from the U.S. Pacific Command pointed out that in their region the level of technical ability varies from country to country and can also vary significantly within countries. In these environments, program success will depend not only on the tools selected, but also on how well principles of sustainability are integrated from the outset of program development and implementation. In some cases, sustainability can hinge on something as basic as equipment maintenance. The original DOD CTR legislation had a "Buy American" provision, which in some cases worked against program sustainability and long-term security impact, especially where the partner country had no local source for regular equipment maintenance and repair of U.S.-origin technology. Projects that incorporate local equipment and technology may have had a greater degree of success. This became the approach that the Department of Energy (DOE) used successfully in its Russian nuclear material protection, control, and

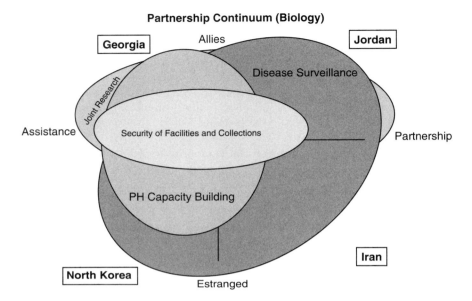

FIGURE 3.1 Partnership Continuum (Biology)
SOURCE: David R. Franz.

accounting program. Not only have local technologies been used, the program has resulted in several spin-off companies that provide security equipment and design security installations.

Figure 3.1 illustrates one way to analyze potential engagements, using biology as an example. Countries on the north-south axis range from allies on the top that are on good terms with the United States and perceive threat in similar ways to estranged countries on the bottom that have difficult or no formal relations with the United States and disagree with the United States on threat perceptions. The east-west axis runs from the countries that require assistance to carry out programs to the countries with their own resources. Based on this analysis, the upper right quadrant offers the richest opportunities for engagement, but at least some level of activity can be projected for all quadrants. If the figure is used to map a biosecurity strategy, it shows that disease surveillance activities can be pursued with almost any country, whereas more sensitive areas like the security of biological facilities and pathogen collections and engaging in joint research are reserved to a more select group of partners.

Finding 3-3: CTR 2.0 will have to tailor approaches for each new engagement and associated threat, and use creative forms of collaboration, particularly in

environments where the partners are reluctant, the political climate is adverse, or local conditions can only support limited levels of technology.

Engagement Strategies

CTR 2.0 programs must be guided by a clear strategy that includes shared responsibility with partner countries for program development, planning, resources, and implementation. This approach should produce a high level of trust and transparency, and promote sustainability. This may sound straightforward, but it will require a leap of faith on the part of U.S. program implementers, who may be more used to "checkbook diplomacy" than true partnership—a "we pay, you do as we say" attitude. Although some CTR 1.0 programs are moving away from this model, the transition to a new, more collaborative model needs to occur quickly.

Nontraditional partners may be able to play important reinforcing or even primary roles. Flexible NGOs or even other countries may be needed to take a lead role in certain circumstances. For example, an NGO partner may have long-term goals for a country or region and be able to maintain a low to moderate level of engagement for an extended period. Partnering may offer CTR 2.0 new opportunities for both sustaining program progress as well as monitoring ongoing implementation once responsibility is assumed by the partner country or countries for sustaining the activity.

> Just because a country may be hesitant to engage in the first instance with a U.S. government program is not necessarily a signal that it will always oppose such engagement; it may just need to be engaged initially in a more creative and limited way.

The challenges to launching a new security engagement may be significant. The committee is aware, for example, of the situation in one country named by several experts as a logical candidate for CTR 2.0 engagement where officials have communicated informally that they are not prepared to discuss USG CTR activities in the nuclear area. However, a USG CTR biosecurity engagement program has established a successful program based on a modest science cooperation program started by the U.S. Agency for International Development. Just because a country may be hesitant to engage in the first instance with a government program is not necessarily a signal that it will always oppose such engagement; it may just need to be engaged initially in a more creative and limited way. The innovative use of a variety of partners could facilitate these early engagement approaches. The "soft engagement" strategy of working in tandem with nongovernment partners will be an important element in future program development.

The broader group of CTR 2.0 partners can help establish initial contacts and relationships in environments where government or international programs, such as support for UNSCR 1540, are desirable but not welcome or feasible in the near term. These facilitators can be the wedge in some circumstances that will pave the way for government programs to follow at a future time, or, in some cases, may have to play a long-term role. This type of soft engagement could involve many different activities, such as training programs, opportunities to participate in professional meetings with individuals or organizations that could be relevant to future efforts, or developing Internet-based networks as a way of initiating dialogue on topics of interest, to name a few. CTR 2.0, therefore, will involve national and international coordination, possible government and nongovernment components, and activity in new regions, with all these elements influencing the shape and content of new engagement strategies.

Finding 3-4: Strategies that employ soft engagement, sometimes facilitated by NGOs, academe, or other nontraditional diplomatic efforts, may be necessary to support or initiate CTR 2.0 engagements.

Recommendation 3-1a: Domestically, CTR 2.0 should include a broad group of participants, including government, academe, industry, nongovernment organizations and individuals, and an expanded set of tools, developed and shared across the U.S. government.

Transparency will be a natural result of CTR 2.0, but the United States must be prepared to accept two-way transparency. One CTR 1.0 program has always had this element because of the way it was initially designed. The Science and Technology Centers (STC) program to redirect the former Soviet WMD scientists and engineers always had an international headquarters staff drawn from all countries, including the host countries, Russia and Ukraine. Although the agreements establishing the STCs require transparency in terms of access to the facilities where projects are funded and program audits, it is really the direct staff involvement that has had a lasting impact. The STC staffs participate in all levels of program implementation, providing significant transparency into operations and management. Because annual project and institutional audits are the norm, all staff members have learned to appreciate the value of oversight and accountability. As CTR 2.0 programs are developed, ways to design transparency into program plans and implementation need to be a priority.

Finding 3-5: Transparency will be a hallmark of CTR 2.0 and will further strengthen commitments to threat reduction beyond any applicable legal obligations in a treaty, contract, or other legal instrument.

New Approaches to Security

The United States and other nations that share a common view of threats have demonstrated that working together to develop innovative approaches can reduce threats. Several efforts have emerged that operate in parallel with traditional arms control treaties. The Proliferation Security Initiative (PSI)[11] operates globally and grew out of the December 2002 U.S. National Strategy to Combat Weapons of Mass Destruction. PSI aims to interdict shipments of WMD, their delivery systems, or materials. The GICNT[12] developed from a joint statement on July 15, 2006, by Presidents George W. Bush and Vladimir V. Putin. It is designed to "expand and accelerate the development of partnership capabilities to prevent, detect, and respond to the global threat of nuclear terrorism."[13] As of July 2008, 75 countries had signed on to the GICNT principles,[14] including some that were named to the committee as possible CTR 2.0 engagement partners.

Another opportunity for CTR 2.0 to support a new international security instrument is the potential for supporting the implementation of UNSCR 1540 and subsequent related resolutions.[15] UNSCR 1540 requires states "to refrain from providing any form of support to non-State actors that attempt to develop, acquire, manufacture, possess, transport, transfer or use nuclear, chemical or biological weapons and their means of delivery."[16] The binding obligations of the resolution include a requirement that states "adopt and enforce appropriate effective laws which prohibit any non-State actor to manufacture, acquire, possess, develop, transport, transfer or use nuclear, chemical, or biological weapons and their means of delivery," and a requirement that states "take and enforce effective measures to establish domestic controls to prevent the proliferation of nuclear, chemical or biological weapons and their means of delivery." It also encourages international cooperation and has a mechanism that allows states to request assistance. The types of assistance under UNSCR 1540 include areas that would be appropriate for CTR 2.0 activities:[17]

---

[11] Department of State. Proliferation Security Initiative. Available as of March 2009 at http://www.state.gov/t/isn/c10390.htm.

[12] State Department. 2006. Global Initiative to Combat Nuclear Terrorism: U.S. Russia Joint Statement. St. Petersburg. July 15. Available as of March 2009 at http://www.state.gov/t/isn/c18406.htm.

[13] Ibid.

[14] See the current list at State Department. 2008. Global Initiatve Current Partner Nations. Available as of March 2009 at http://2001-2009.state.gov/t/isn/105955.htm.

[15] See United Nations Security Council. Resolutions. 2004. Available as of March 2009 at http://www.un.org/Docs/sc/unsc_resolutions04.html for the text of the resolution. See also UNSCR 1673 (2006) and UNSCR 1810 (2006), available as of March 2009 at http://www.un.org/Docs/sc/unsc_resolutions06.htm. See also UNSCR 1810 (2008) available as of March 2009 at http://www.un.org/Docs/sc/unsc_resolutions08.htm.

[16] Ibid.

[17] Ibid.

- Drafting and implementing legislation relevant to prohibiting state support to nonstate actors that attempt to acquire WMD or means of delivery or to conduct any other activity set forth in operative paragraph (OP) 1
- Drafting and implementing legislation to prohibit nonstate actors from conducting, attempting to conduct, participating as an accomplice, assisting, or financing any activity set forth in OP 2 relating to WMD or means of delivery
- Developing and implementing measures to account for and secure related materials to prevent the proliferation of WMD or means of delivery as set forth in OP 3(a). *Note: Assistance could include equipment or training relating to the development of measures.*
- Developing and implementing physical protection measures to prevent the proliferation of WMD or means of delivery as set forth in OP 3(b). *Note: Assistance could include equipment or training relating to the development of measures.*
- Developing and implementing measures, equipment, or training to improve border controls and law enforcement efforts as set forth in OP 3(c). *Note: Assistance could include equipment or training relating to the development of measures.*
- Establishing and maintaining effective national export and transshipment controls over WMD or means of delivery and related materials, as well as controls on providing funds and services relating to such export and transshipment, such as financing, including the drafting or improvement of relevant legislation as set forth in OP 3d
- Drafting, updating, or implementing lists of export controlled items as set forth in OP 6
- Developing appropriate ways to work with and inform industry and the public regarding their obligations as set forth in OP 8(d)
- Other requests (e.g., for demonstrations of equipment, technical briefings, informal consultations, and so on)

In this regard, the DOD CTR WMD–Proliferation Prevention Initiative (PPI) is well positioned to respond to some of these needs based on the work it has done to prevent proliferation of WMD across the borders of non-Russian states in Eurasia. The PPI mission of countering efforts by terrorists to secure WMD and WMD components, materials, and expertise is consistent with UNSCR 1540; and the program goals of improving the security of states' borders, building the capacity of states to investigate WMD-related thefts and smuggling, and securing any WMD materials within their borders are either directly relevant to other environments or could be modified to respond to those environments. Current program activities include providing equipment, logistics support, training, and other support to appropriate partner country government agencies. The PPI has worked with departments of defense, departments of interior, national guards, border guards, and customs organizations in partner countries, and has coordinated its efforts with related U.S. programs in the Departments of State, Energy, and Commerce, and the U.S. Coast Guard.

In addition, the DOD CTR Defense and Military Contacts Program already carries out activities that could be applied easily to a global environment. Specific suggestions for how this program could be applied by DOD CTR as an element of CTR 2.0 are in Chapter 4.

Some countries have already submitted requests to the UNSCR 1540 Committee for assistance (see Table 3.1), but others could be encouraged to do so as part of a CTR 2.0 UNSCR 1540 initiative. Part of the incentive for new partner countries to cooperate on UNSCR 1540 implementation would be to establish their credentials as responsible members of the international community, but engagement could also help them solve internal or transborder issues. For some potential partners, 1540 compliance is low on the list of priorities compared to other more immediate challenges of providing food, water, and shelter for their citizens. At the same time, however, some of these countries recognize that engaging in a partnership activity, even if they can only add limited resources as their contribution to the partnership, may be to their advantage. In discussions with officials at U.S. African Command, committee members were told that UNSCR 1540 implementation may be a low priority in Africa, but combating smuggling and trafficking is urgent. Framing UNSCR 1540 in a way that addresses national and regional issues could open new opportunities. Senior officers at U.S. African Command commented that most of the countries in their region would be difficult to engage in border control or customs assistance, but might respond positively if the same set of programs is presented as countersmuggling assistance, which is a priority throughout the region.

Engagement can be initiated both through diplomatic channels led by the Department of State or through broader multilateral or international contacts or both. Developing CTR 2.0 programs in this way could have several advantages. First, for countries that may be sensitive about working bilaterally with the United States, using an international approach could ease tensions. Second, we know from other nonproliferation programs that it is easier to diversify funding when the program base is multilateral or international. The STCs are a good example of this, as are certain activities under the International Atomic Energy Agency technical assistance program. In addition, the experience from the G8 GP shows that countries with small amounts of funding to contribute often can do so only if there is a mechanism that allows them to pool their funds or to piggyback funding, allowing the small donor to transfer its funds to a larger donor that may already have a program agreement in place. G8 GP experts commented to the committee that they are convinced that small donors will be lost if facilities are not available to piggyback funds and that there already are

TABLE 3.1 UNSCR 1540: Requested CTR-Relevant Assistance by Category and Country

	Legislation and Regulatory Assistance	Import/Export Licensing and Controls	Border or Customs Assistance	Detection Training or Detection Hardware	Police, Border, Customs, Military Training	Generally Interested
Albania	X	X	X			
Angola						X
Armenia						X
Bahamas	X		X	X	X	
Belize						X
Benin	X		X			
Bolivia						X
Cambodia	X			X	X	
Colombia	X		X	X	X	
Guatemala				X	X	
Jamaica						X
Lebanon	X					X
Lithuania				X		
Marshall Islands		X	X	X		X
Morocco	X				X	X
Philippines	X		X	X	X	ports
Serbia	X		X	X	X	
Syria						X
Thailand	X	X		X	X	
Uzbekistan	X	X			X	

requests to set up a common fund at the UN to support 1540 implementation. (See also "Funding with International Partners" later in this chapter.)

Finding 3-6: In addition to supporting traditional arms control and nonproliferation agreements, CTR 2.0 can be used to advance other multilateral (Proliferation Security Initiative, Global Initiative to Combat Nuclear Terrorism) and various international security instruments such as UNSCR 1540 and related resolutions.

CTR 2.0 IN POST-CONFLICT ENVIRONMENTS

Post-conflict environments may offer a particular opportunity to include CTR 2.0 as countries seek ways to reestablish security. In Iraq, in parallel with ongoing military action, several USG CTR programs were initiated in late 2003 to address potential threats. DOE experts and U.S. military forces worked together with Iraqis to gather and safeguard radiological sources; chemical munitions were stored in secure facilities, and the State Department supported export control and border security training, and a program to redirect former Iraqi weapons scientists. These programs demonstrate that even in adverse environments, it is possible to engage in constructive ways.

CTR 2.0 should develop lighter, more responsive, more agile, and more easily deployed tools, and through this flexibility and adaptability it can both reduce the proliferation of WMD as well as take on more soft engagement tasks in post-conflict periods. Multilateral coalitions of militaries, and when possible, NGOs, industry, and even academe, have tools that can be applied effectively in the transition from war to peace that also address important security vulnerabilities. The principles of relevance to the needs of the people of a region, building toward sustainment and moving from patronage to partnership apply in this role for CTR 2.0 as they do in the traditional role for reducing proliferation of WMD. Once understood, established, and codified as a tool of foreign policy, CTR 2.0 can be a very powerful tool to be used alongside the other tools of national security policy that can be applied in many situations.

In this context, finding a role for the Unified Combatant Commands in CTR 2.0 could produce some important and innovative approaches to post-conflict engagement. These could include encouraging countries to comply with international arms control treaties, UNSCR 1540 implementation, and participation in other similar efforts. Although these may not be a country's highest post-conflict priorities, they are important to the international community, particularly if the country possesses any technical or WMD capability that could be vulnerable in a post-conflict environment. For example, if a country has any kind of chemical or biotechnology research or industrial capacity, programs to improve physical security, biological and chemical laboratory safety and security, export controls, and border security may be relevant. Likewise, reestablishing health care and public health capabilities as quickly as possible will help the populations feel safer and more secure, and also contribute to an alert network to detect natural or deliberate disease outbreaks. And as Iraq has taught us, countries with unsecured munitions can be both the source and the target of improvised explosive devices. Addressing these issues in the immediate post-conflict period may help prevent them from becoming a new threat later.

Finding 3-7: The holistic approach of CTR 2.0, including engagement with international partners, can be useful in post-conflict environments.

Recommendation 3-1b: Internationally, CTR 2.0 should include multilateral partnerships that address both country and region-specific security challenges, as well as provide support to the implementation of international treaties and other security instruments aimed at reducing threat, such as the G8 Global Partnership, the Proliferation Security Initiative, UNSCR 1540, and the Global Initiative to Combat Nuclear Terrorism.

PERSONAL ENGAGEMENT

Professional colleagues—friend or foe—throughout the world respect intellect and technical competence. Relationships provide opportunities for communication, access, and even transparency in times of great national tension, and may be one of the most important achievements of CTR programs. From the early DOD CTR senior-level military exchanges to recent collaborations in disease surveillance, close relationships formed around professional interactions persist, even where tensions between countries are heightened. Because of the fundamental change in the nature of threats and the pace at which events occur, the ability to communicate directly with a specialist in another country on a regular basis—to discuss an emerging disease with a fellow public health official or a terrorist attack in his or her country—has greater national security significance today than it did when CTR was founded.

Even after specific cooperative threat reduction efforts in Russia and the FSU are completed, valuable personal relationships between individuals—particularly between scientists, engineers, and military officers and government officials—will remain, providing continued opportunities for communication and even informal transparency. These personal relationships have become the foundation for further professional or technical collaborations in some cases. For example, contacts made during the Intermediate-Range Nuclear Forces Treaty negotiations in the 1980s provided opportunities later when the same people worked together in other cooperative environments, even after retirement from official government service. More recent cooperative studies that examine mutual problems and barriers to continuing progress have reinforced the value of such relationships. These professional associations exist at many levels, and the continuing contacts provide not only channels for future communication but also a way to judge more clearly the meanings of political statements, actions taken, and pronouncements made during periods of international tension. In the early post-Cold War years, contacts made through CTR initiatives were invaluable to gauge political, social, and security developments within the FSU and vice versa. They also helped build the trust necessary to secure access by both countries to formerly sensitive sites. CTR 2.0 should encourage and expand such ties.

Building relationships between professional colleagues will be facilitated by personnel stability on the U.S. side. DOE experience in Russia suggests that our partners value seeing familiar faces over time and that this is an important element of building trust. When sensitive facilities and technologies are involved, the importance of continuity cannot be understated. Similarly, having in-country contacts resident in embassies also helps form strong program links, promotes a sense of teamwork and collaboration, and can manage questions before they develop into problems.

Several individuals highlighted to the committee the constructive role that scientist-to-scientist and collaborative technical relationships can have. These were suggested as an excellent way to establish new relationships that can provide low-key cooperative continuity while governments explore broader cooperation. This kind of collaboration can also provide insights into whether a partner's technological abilities really represent a threat or not. Historically, the Departments of State and Energy have made excellent use of this approach in their CTR programs. As noted in the discussion of metrics (Chapter 2), ways to reflect personal engagement as a factor of program success need to be developed.

Finding 3-8: The benefits of "personal engagement" survive beyond the formal implementation of programs and projects.

Recommendation 3-2: The executive branch and Congress need to recognize that personal relationships and professional networks that are developed through USG CTR programs contribute directly to our national security and that new metrics should be developed to reflect this.

CTR 1.0 TO CTR 2.0: THE GLOBAL SECURITY CONTINUUM

Congress has done much over the years to amend legislation in ways that allow USG CTR programs to work more broadly and effectively, but some legal and policy underpinnings of the current CTR 1.0 program are cumbersome and dated and often diminish the value of assistance and partnership programs. Although the DOD CTR authorizing legislation has undergone some fundamental, positive changes, several issues need to be addressed to allow CTR 2.0 to operate optimally. Some of these may require congressional action; others may require executive branch action.

> In all respects, the committee's observations about the need for leadership, coordination, and cooperation in the executive branch apply equally to Congress.

The committee believes that accomplishing the necessary changes will require regular consultation between the legislative and executive branches. Senators Sam Nunn and Richard Lugar have been strong and vocal champions of CTR 1.0, and without their vision and commitment the program would not exist. But CTR 2.0 is an even more complex and possibly larger endeavor; it, too, will require champions and a forum in which both they and critics can discuss the many issues that will arise. The committee's observations about the need for leadership, coordination, and cooperation in the executive branch apply equally to Congress.

The legislative changes recommended by the committee are limited to significant impediments to establishing a functioning CTR 2.0 framework. The recommendations are narrowly targeted on issues that were raised repeatedly during the course of the committee's consultations with experts.

Recommendation 3-3: The legislative framework, funding mechanisms, and program leveraging opportunities should be structured to support more effective threat reduction initiatives across DOD, other U.S. government departments and agencies, international partners, and NGOs.

The CTR 2.0 Budget

Many experts commented on budget-related issues to the committee. It became clear that just as there is a gap in USG CTR program strategy and coordination, there is a similar gap on the budget side. There is no White House or Office of Management and Budget (OMB) effort to request and align funding in a way that takes into account needs, capabilities, or priorities. Some of these budget problems can perhaps be remedied by making the right officials aware of the problem; others will need longer-term, more complex solutions and changes of policy by OMB. Some of the issues that were raised and suggestions on possible approaches follow.

- Aligning budget periods to match strategic planning and programming: Funds are appropriated to different U.S. government agencies for different durations. One agency involved in USG CTR programs may receive funds that have to be obligated within the fiscal year in which they were appropriated;

another may receive funds that can be used over 3 fiscal years. Adding to the problem, OMB generally does not support a multiyear budget process.
- Recognizing the broad set of departments and programs that contribute to USG CTR: Not all departments and programs that support CTR efforts currently have the appropriate national security legal authorities that clearly identify them as USG CTR participants. "Nonsecurity agencies," such as the Departments of Health and Human Services (HHS) and Agriculture, and the Environmental Protection Agency, have become indispensable partners in CTR 1.0 work, but this is not reflected in legislative authorities or appropriations. To manage the budget and program coordination problems this situation causes, the national security agencies (Defense, State, or Energy), whose authorities specifically include national security missions, to transfer funds to nonsecurity agencies. This is a very inefficient approach. The committee was told about a situation that exists between the State Department and HHS that illustrates this point well.

  o The Department of State CTR office often receives its budget allocation from the Treasury Department late in the fiscal year. At the point the funding is received, the State Department develops interagency transfer agreements with other agencies, such as HHS, which provide essential technical expertise for the State Department's biosecurity engagement programs. By the time the funds are transferred to HHS, the fiscal year is typically drawing to a close. The State Department appropriations may be spent over 3 years, but the HHS appropriation has to be spent in 1 year, and its accounts are only structured for 1-year money. By the time State Department funds reach HHS and are put in a 1-year account, there is no time to allocate the funds to programs. This same mismatch scenario is played out between other agencies as well, making it difficult to plan or implement any program on a rational basis. Program delays caused by this situation also lead to confusion and misunderstandings with international partners.

Possible Approaches: Two possible approaches to address these issues illustrate why it is important to involve OMB and a broad cross section of Congress in these discussions.

  1. Preferred Approach: The executive branch works with Congress to add authorities to the departments whose participation is crucial to the success of global security engagement programs. These authorities can then be matched by appropriations directly to the departments instead of providing them through other agencies. Part of this process should include designating CTR 2.0 funds as multiyear to ensure program flexibility.
  2. Interim Approach: Since establishing new authorities may take some time, OMB and the agency in question can work with the congressional appro-

priators to ensure that the agency receiving funds has the flexibility to create appropriate accounts to accept other agency funds.

The preferred approach leads to a larger issue of how much funding should be allocated, and to which budgets, to support critical security engagement work. At present, each agency develops its own budget, which goes through a stove-piped process to OMB, where budgets are adjusted to meet a maximum presidential budget figure for any given fiscal year. Gordon Adams, a former senior official at OMB, describes this as "the diaspora of foreign assistance programs."[18] In addition, the 2009 DOD Quadrennial Roles and Missions Review "supports institutionalizing whole-of-government approaches to addressing national security challenges," including the budgets of national security programs.[19] In the current system, there is no referee at the White House level looking across the many agencies and programs that could contribute to CTR 2.0 to determine if adequate resources are going to the programs best able to accomplish the priority tasks that have been defined in the White House-led strategic planning exercise.

Recommendation 3-3a: Program planning should be developed out of a strategic process and be matched by a *strategic budget process* that produces a multiyear budget plan and distributes funding across agencies based on agency ability to respond to program requirements. As needed, agency legislative authorities should be revised to include a national security dimension.

Funding with International Partners

The congressional request for this study expressed a particular interest in how USG CTR programs can work more effectively with international partners and how, through those partnerships, the United States can encourage more partner funding. In reviewing this question, the committee determined that the current lack of comingling authority needs to be addressed.

The Miscellaneous Receipts Act requires that money received by the U.S. government be deposited into the General Fund of the U.S. Treasury. The act was passed to ensure that, as a general matter, government agencies do not bypass the appropriations authority of Congress by augmenting their budgets

---

[18] Gordon Adams. 2008. Smart Power: Rebalancing the Foreign Policy/National Security Toolkit. Testimony before the Subcommittee on Oversight of Government Management, the Federal Workforce, and the District of Columbia of the Senate Committee on Homeland Security and Governmental Affairs. Hearing on a Reliance on Smart Power–Reforming the Foreign Assistance Bureaucracy. July 31.

[19] Department of Defense. 2009. *Quadrennial Roles and Missions Review Report.* Washington, D.C.: Department of Defense. 31 pp. Available as of March 2009 at http://www.defenselink.mil/news/Jan2009/QRMFinalReport_v26Jan.pdf.

via other means. The Carter-Joseph Report urged Congress to exempt DOD CTR from the Miscellaneous Receipts Act[20] by authorizing DOD CTR to accept funds from foreign countries and to comingle those with appropriate DOD CTR funds. This would enable countries, for example, to contribute to DOD CTR in fulfillment of their G8 GP commitments without having to negotiate their own separate umbrella agreements. Such comingling authority exists broadly in other countries, such as the United Kingdom and Canada, and has been provided by Congress for some specific DOE programs, including Second Line of Defense[21] and Global Threat Reduction Initiative. This issue was raised with several G8 GP partner countries, who argued that this ability was critical for securing the contributions of small donors who otherwise would not apply their funds to CTR-type programs. The committee believes that if Congress provides all agencies operating under CTR 2.0 with such comingling authority for CTR purposes, it will increase the potential for countries to share in program costs. Having this authority would also help address the issue of differing authorities, budgets, and time lines of international partners. The case frequently arises where a country's desire to contribute to a project does not mesh with its legal and budgetary structures. Comingling authority adds the additional flexibility that may make participation possible in such cases.

Recommendation 3-3b: Congress should provide *comingling authority* to all agencies implementing programs under CTR 2.0 as a way to encourage other partners to contribute funds to global security engagement efforts.

Legal Frameworks

U.S. government programs have adopted a variety of legal frameworks under which CTR 1.0 has been implemented. The committee believes that implementation of the DOD CTR program is hindered by the relative lack of flexibility in its legal frameworks and authorities. These include the following:

- umbrella agreement issues relating to liability, taxes, and access
- geographic limitations and burdensome contracting procedures that could be eased by the provision of "notwithstanding authority"
- the lack of "comingling authority"

---

[20] Miscellaneous Receipts Act, 31 U.S.C. § 3302(b)(2006). Available as of March 2009 at http://frwebgate3.acces.gpo.gov/cgi-bin/TEXTgate.cgi?WAISdocID=01862318241+0+1+0&WAISaction=retrieve.

[21] National Nuclear Security Administration (NNSA). 2008. NNSA's Second Line of Defense Program. Department of Energy. Available as of March 2009 at http://nnsa.energy.gov/news/992.htm.

Finding 3-9: Many of the legal and policy underpinnings of the current DOD CTR program that were established for accountability and protection of U.S. implementing agencies are cumbersome, dated, and limiting, and often diminish the value and hinder the success of program assistance and partnerships.

The DOD CTR Umbrella Agreement: Issues Relating to Liability, Access, and Taxation

The DOD CTR bilateral umbrella agreement is well established as the mechanism under which programs are implemented. Its provisions have changed little over time, and recent experiences, including multiyear negotiations to extend the Russian umbrella agreement and to establish an umbrella agreement with Kazakhstan, signal that it may be time to consider other approaches.

On June 19, 2006, the United States and Russia signed a protocol to extend for another 7-year period the U.S.-Russian Cooperative Threat Reduction Umbrella Agreement, which entered into force in 1992, and was first extended in 1999. As a result of protracted negotiations over the agreement's liability protections, the 2006 Extension Protocol was signed less than a week before the agreement was due to expire.[22] Press reports portrayed the DOD CTR program as nearly derailed by the dispute.[23] The DOD CTR agreement's access and taxation exemption provisions have also been the subject of contention. Disputes over liability, access, or taxation could again threaten the umbrella agreement's extension when the 2006 protocol expires in 2013.

Liability and access issues in particular could also hinder progress in the interim. The 2006 DOD CTR extension protocol kept the original umbrella agreement liability protections in place for existing projects, but left protection language for future projects subject to negotiation.[24]

The access provisions of the DOD CTR umbrella and related agreements provide the U.S. government the right to examine the use of materials or services provided by it as part of the assistance process. However, a February 2007 report by the Government Accountability Office (GAO) warned of continuing restrictions on U.S. access to facilities that store, manufacture, or dismantle

---

[22] Peter Baker. 2006. U.S., Russia Break Impasse on Plan to Keep Arms from Rogue Users. *Washington Post*. June 20. A11 pp.

[23] Peter Eisler. 2006. U.S., Russia reach deal on securing Soviet WMD; Post-Cold War program nearly derailed by dispute. *USA Today*. June 16. See also, Michael Crowley. 2007. The Stuff Sam Nunn's Nightmares Are Made Of. *New York Times*. February 25. The two sides signed an agreement to move ahead with plutonium disposition in 2000, but the deal could not be implemented until a liability protocol was signed some 5 years later. See U.S., Russian Officials Sign Liability Protocol for Plutonium Disposition. *Inside the Pentagon*. September 21, 2006.

[24] Eisler.

Russian nuclear weapons.[25] The GAO report noted that "access difficulties at some Russian nuclear warhead sites may . . . prohibit DOE and DOD from ensuring that U.S.-funded security upgrades are being properly sustained."[26] For example, "Russia has denied DOE access at some sites after the completion of security upgrades, making it difficult for the department to ensure that funds intended for sustainability of U.S.-funded upgrades are being properly spent."[27] Specifically, neither DOE nor DOD had "reached an agreement with the Russian [Ministry of Defense] on access procedures for sustainability visits to 44 permanent warhead storage sites where the agencies are installing security upgrades."[28] Absent such agreement, DOE and DOD "will be unable to determine if U.S.-funded security upgrades are being properly sustained and may not be able to spend funds allotted for these efforts."[29] Such limitations could impede compliance with U.S. laws requiring verification of the proper use of U.S. government funds.

Perhaps the best-known standoff over access involves the DOD-funded Fissile Material Storage Facility at Mayak.[30] The United States and Russia from the outset of the project agreed in principle that the United States would have the right to some form of monitoring of this site, to ensure that it is being used for its intended purpose. However, 5 years after the site was commissioned and 10 years after transparency negotiations began, a transparency agreement has not been concluded.

The umbrella agreement issues have been the subject of tensions not only between the United States and Russia but also between U.S. departments and agencies. As the State and Energy departments began to join the Defense Department in funding and implementing CTR-type projects, the former Soviet states and especially Russia learned to play U.S. agencies off each other—seeking weaker legal protections from one U.S. department and then arguing the new provisions served as a precedent for other U.S. departments.

> Arguably, if it takes 2 or more years to put an umbrella agreement in place before any work begins, the nature and urgency of the threat being addressed has to be questioned.

---

[25] GAO. 2007. *Progress Made in Improving Security at Russian Nuclear Sites, but the Long-term Sustainability of U.S.-Funded Security Upgrades is Uncertain.* Available as of March 2009 at http://www.gao.gov/new.items/d07404.pdf.
[26] Ibid., p. 22.
[27] Ibid., p. 26.
[28] Ibid., p. 29.
[29] Ibid.
[30] Matthew Bunn. 2007. Securing Nuclear Warheads and Materials: Mayak Fissile Materials Storage Facility. Nuclear Threat Initiative. Available as of March 2009 at http://www.nti.org/e_research/cnwm/securing/mayak.asp.

As we look forward to broadening engagements, it is time to look carefully at what mechanisms are required and how they should be applied. If CTR 2.0 programs are to form a meaningful response to situations that pose a threat to U.S. national security, implementation mechanisms will have to be put in place in a timely manner. Arguably, if it takes 2 or more years to put an umbrella agreement in place before any work begins, the nature and urgency of the threat being addressed has to be questioned. As a result of the 2008 G8 Summit, the G8 GP has accepted the principle of expanding beyond the former Soviet Union.[31] The G8 GP already has a set of guidelines for new programs, and the committee was informed that an effort may be under way to develop a model G8 GP project agreement. If basic model project agreement terms could be articulated, this might help accelerate the process of putting new agreements in place. Other rapid contracting mechanisms also should be explored.

DOD would benefit from undertaking a systematic study of its umbrella agreement and other contracting mechanisms. It needs to identify where the DOD CTR program is currently prohibited by law from starting work and which specific contracting procedures may be responsible for the DOD CTR program's inability to move with requisite speed and efficiency. It is better that these obstacles be identified now, and if appropriate, removed quickly, rather than be identified at a time when the provisions stand in the way of accomplishing a high-priority national security goal. This will provide needed CTR 2.0 program flexibility and allow programs to respond to important opportunities that may be lost while waiver authority is sought.

Recommendation 3-3c: To maximize the effectiveness of CTR 2.0, the DOD CTR legal frameworks and authorities should be reassessed. DOD should undertake a systematic study of the *CTR Umbrella Agreement* protection provisions, what purposes they serve in particular circumstances, whether there might be less intrusive means of accomplishing the provisions' goals, and when the provisions are necessary in their present form. In addition, all USG CTR programs should identify legal and policy tools that can promote the sustainability of U.S.-funded CTR work and provide greater implementation flexibility.

Geographic Limitations, Contracting Procedures, and "Notwithstanding Authority"

> As a practical matter, the State Department's Nonproliferation and Disarmament Fund largely operates in the absence of government-to-government liability, taxation, and access protection provisions. It relies instead on mechanisms such as contracts with its foreign counterparts, and asserts that its diminished protections have not led to problems.

---

[31] See Appendix G.

This committee agrees with individuals at DOD and elsewhere who have suggested that the traditional DOD CTR Umbrella Agreement may not be necessary for some countries to which DOD might expand. Depending on the anticipated scope of work, this is undoubtedly correct as a matter of law. As a practical matter, the State Department's Nonproliferation and Disarmament Fund (NDF) program largely operates in the absence of government-to-government liability, taxation, and access protection provisions. It relies instead on mechanisms such as contracts with its foreign counterparts, and asserts that its diminished protections have not led to problems. NDF and its flexible structure will be an important element of CTR 2.0.

Because of the difficulty of negotiating and extending traditional DOD CTR Umbrella Agreements and NDF's success in operating in their absence, a study that looks at the two models could contribute significantly to enabling DOD CTR to operate more nimbly. DOD (and other U.S. government agencies) could also study other existing arrangements between the United States and potential partner countries, such as science and technology, health, or other agreements, to assess whether these might provide an adequate framework, particularly for any initial engagement work.

While the NDF receives its funding from Congress for expenditure "notwithstanding any other provision of law," DOD CTR has no such notwithstanding authority. As a result, DOD CTR is subject to geographic limitations, contracting procedures, and other restrictions that do not apply to NDF.

Geographic Limitations

Beginning with the Fiscal Year 2004 Defense Authorization Act, Congress began authorizing the President to use a portion of DOD CTR funds outside the former Soviet Union in emergency situations. The George W. Bush administration exercised this authority for the first time in mid-2004, when it provided assistance to Albania for the elimination of chemical weapons. In 2007, Congress expanded the authority to spend DOD CTR funds outside the FSU by eliminating the restriction that this occurs only in emergency situations. However, the program is still subject to the Glenn Amendment[32] and other similar sanctions, which could be an obstacle to work in countries subject to those sanctions.[33] The Supplemental Appropriations Act of 2008 provided the President with Glenn Amendment waiver authority with respect to CTR-type work

---

[32] The "Glenn Amendment," or the Nuclear Proliferation Prevention Act of 1994, imposes sanctions under several conditions, including on nonnuclear states that detonate nuclear explosions. See also the Glenn-Symington Amendments to the Foreign Assistance Act in 1977 and the Nuclear Nonproliferation Act of 1978.

[33] For many years, Congress conditioned funding for the Cooperative Threat Reduction Program on the president making an annual certification that each recipient nation was "committed to" certain goals. However, in 2007, Congress eliminated the certification requirements.

in the Democratic People's Republic of Korea (DPRK). The Glenn Amendment was lifted with respect to India and Pakistan shortly after September 11, 2001. Sources with whom committee members spoke disagreed as to whether DOD CTR work in Iran or any other country is currently barred by a federal law or laws that cannot be waived by the president. However, existing waiver authorities do not take into account the potential for future sanctions that may not be subject to waiver. For example, the list of State Sponsors of Terrorism can be revised, and any country added to that list would be subject to sanctions. In such cases, it may not be possible to engage under any program other than the NDF. Given NDF's relatively small annual appropriation, it is possible that the bulk of its funds might be used by a single program (such as denuclearization in the DPRK), leaving no backup program with similar flexibility to take on a new activity. There also can be specific prohibitions contained in appropriations language, as is the case for DOD and the DPRK.

Contracting Procedures

Unlike NDF, DOD CTR is subject to the Federal Acquisition Regulations and other federal contracting procedures and restrictions. Several sources opined to the committee that these requirements were a major reason why DOD CTR is sometimes unable to match NDF's speed and lower cost estimates.

"Notwithstanding" Authority

The geographic limitation and contracting issues can be addressed through limited provision of notwithstanding authority. Senator Lugar has proposed that the DOD CTR program be given authority to act "notwithstanding" any sanction or other provision of law, to ensure that the program would have the ability to respond rapidly to new nonproliferation opportunities. The Carter-Joseph Report also recommended that Congress provide DOD CTR with notwithstanding authority comparable to that enjoyed by NDF or, failing that, provision for specific waivers in high-priority cases.[34] Although several congressional staff members with whom committee members spoke expressed opposition to providing DOD CTR with blanket notwithstanding authority, the committee believes that limited notwithstanding authority is needed to provide the U.S. government with adequate flexibility. Specific exceptions, such as the

---

[34] Ashton B. Carter, Robert G. Joseph, et al. 2008. *Review Panel on Future Directions for Defense Threat Reduction Agency Missions and Capabilities to Combat Weapons of Mass Destruction*. Cambridge: Harvard University. Available as of March 2009 at http://belfercenter.ksg.harvard.edu/publication/18307/review_panel_on_future_directions_for_defense_threat_reduction_agency_missions_and_capabilities_to_combat_weapons_of_mass_destruction.html?breadcrumb=%2F.

congressional waiver authority provided to the Glenn Amendment for CTR-type work (including by DOE) in the DPRK, are not sufficient and do not take into account the limitations of possible future sanctions.

Recommendation 3-3d: Congress should grant DOD limited "*notwithstanding*" *authority* for the CTR program—perhaps a maximum of 10 percent of the overall annual appropriation and subject to congressional notification—to give the program the additional flexibility it will need in future engagements.

SUMMARY OF CHAPTER FINDINGS AND RECOMMENDATIONS

Finding 3-1: The lack of a government-wide tracking program for USG CTR programs that cross agency budgets impedes the U.S. government's ability to develop a strategic approach to CTR 2.0.

Finding 3-2: Responding to the new global security challenges requires a new model of interagency leadership. CTR 2.0 will function most effectively with strong leadership from the White House, and with the active involvement of relevant departments and agencies.

Recommendation 3-1: CTR 2.0 should be directed by the White House through a senior official at the National Security Council and be implemented by the Departments of Defense, State, Energy, Health and Human Services, and Agriculture, and other relevant cabinet secretaries.

Finding 3-3: CTR 2.0 will have to tailor approaches for each new engagement and associated threat, and use creative forms of collaboration, particularly in environments where the partners are reluctant, the political climate is adverse, or local conditions can only support limited levels of technology.

Finding 3-4: Strategies that employ soft engagement sometimes facilitated by NGOs, academe, or other nontraditional diplomatic efforts, may be necessary to support or initiate CTR 2.0 engagements.

Recommendation 3-1a: Domestically, CTR 2.0 should include a broad group of participants, including government, academe, industry, nongovernmental organizations and individuals, and an expanded set of tools, developed and shared across the U.S. government.

Finding 3-5: Transparency will be a hallmark of CTR 2.0 and will further strengthen commitments to threat reduction beyond any applicable legal obligations in a treaty, contract, or other legal instrument.

Finding 3-6: In addition to supporting traditional arms control and nonproliferation agreements, CTR 2.0 can be used to advance other multilateral (Proliferation Security Initiative, Global Initiative to Combat Nuclear Terrorism) and various international security instruments such as UNSCR 1540 and related resolutions.

Finding 3-7: The holistic approach of CTR 2.0, including engagement with international partners, can be useful in post-conflict environments.

Recommendation 3-1b: Internationally, CTR 2.0 should include multilateral partnerships that address both country and region-specific security challenges, as well as provide support to the implementation of international treaties and other security instruments aimed at reducing threat, such as the G8 Global Partnership, the Proliferation Security Initiative, UNSCR 1540, and the Global Initiative to Combat Nuclear Terrorism.

Finding 3-8: The benefits of personal engagement survive beyond the formal implementation of programs and projects.

Recommendation 3-2: The executive branch and Congress need to recognize that personal relationships and professional networks that are developed through USG CTR programs contribute directly to our national security and that new metrics should be developed to reflect this.

Recommendation 3-3: The legislative framework, funding mechanisms, and program leveraging opportunities should be structured to support more effective threat reduction initiatives across DOD, other U.S. government departments and agencies, international partners, and NGOs.

Recommendation 3-3a: Program planning should be developed out of a strategic process and be matched by a *strategic budget process* that produces a multiyear budget plan and distributes funding across agencies based on agency ability to respond to program requirements. As needed, agency legislative authorities should be revised to include a national security dimension.

Recommendation 3-3b: Congress should provide *comingling authority* to all agencies implementing programs under CTR 2.0 as a way to encourage other partners to contribute funds to global security engagement efforts.

Finding 3-9: Many of the legal and policy underpinnings of the current DOD CTR program that were established for accountability and protection of U.S. implementing agencies are cumbersome, dated, and limiting, and often diminish the value and hinder the success of program assistance and partnerships.

Recommendation 3-3c: To maximize the effectiveness of CTR 2.0, the DOD CTR legal frameworks and authorities should be reassessed. DOD should undertake a systematic study of the *CTR Umbrella Agreement* protection provisions, what purposes they serve in particular circumstances, whether there might be less intrusive means of accomplishing the provisions' goals, and when the provisions are necessary in their present form. In addition, all USG CTR programs should identify legal and policy tools that can promote the sustainability of U.S.-funded CTR work and provide greater implementation flexibility.

Recommendation 3-3d: Congress should grant DOD limited "*notwithstanding*" *authority* for the CTR program perhaps a maximum of 10 percent of the overall annual appropriation and subject to congressional notification—to give the program the additional flexibility it will need in future engagements.

4

The Role of the Department of Defense in Cooperative Threat Reduction 2.0

DRAWING ON ESTABLISHED STRENGTHS

The original Department of Defense (DOD) Cooperative Threat Reduction (CTR) activities in Russia and the former Soviet Union (FSU) focused heavily on military engagement and the destruction and dismantlement of massive weapons systems and the facilities that developed them.[1] CTR 2.0 must address much more complex and diverse security threats. Some CTR 2.0 efforts may be able to take advantage of the original DOD CTR programs, but tasks that require the magnitude of effort needed to address the FSU's weapons of mass destruction (WMD) arsenal are likely to be the exception. DOD is not the only U.S. government department that is capable of conducting CTR activities, nor may it always be the best choice to undertake certain tasks, but it has core strengths that will make it an indispensable part of CTR 2.0.

DOD CTR has significant experience in implementing complex, multiyear projects and can draw on its base of contractor support. In addition, DOD CTR can draw on DOD resources to provide logistics support. With respect to the latter, an important lesson learned from the Libya experience is that the DOD ability to provide a rapid air or sealift response is tempered by other ongoing priority missions. In the committee's discussions about the Libya experience, it learned that it would have taken a fairly senior decision maker to reprioritize an airlift because of pressing logistics requirements in Afghanistan and Iraq. CTR 2.0 must have immediate access to such a decision maker, one who has sufficient knowledge of all requirements to ensure that critical needs are met. Even though DOD is often the logical source for logistical support in these matters, it may not always be able to respond in the time required. CTR 2.0

---

[1] See Appendix I for a list of current DOD CTR programs.

strategic plans, therefore, need to take these limitations into account and plan redundancies accordingly.

The application of DOD CTR to CTR 2.0 may also draw on elements of large CTR 1.0 programs. For example, DOD CTR provided environmental monitoring laboratories (and associated training) related to chemical weapons destruction in Russia and biological weapons facility dismantlement in Kazakhstan. DOD CTR's experience with this kind of project may make it a good candidate for establishing a similar monitoring capability and training program associated with the nuclear dismantlement activities in the Democratic People's Republic of Korea (DPRK) when conditions exist that would permit engagement there.

CTR 2.0 strategic planners will need to measure where DOD CTR will be welcome as a partner and where it will not. Although some countries may appreciate U.S. military involvement, others may view the inclusion of DOD CTR as an attempt to dismantle military assets, particularly in early stages of engagement. For example, the committee learned that conservative elements in India objected to Section 109 of the Hyde Act[2] because it called for the establishment of a CTR program. That was interpreted by some Indians as an attempt to dismantle India's nuclear capability. Even though the title of the section was changed and the intent was to develop nuclear nonproliferation cooperation with the Department of Energy (DOE), suspicions lingered. As stated elsewhere in this report, the committee believes that establishing the initial point of engagement will be a critical step for any CTR 2.0 activity and careful choices must be made about how to launch an effort most effectively.

Finding 4-1: DOD CTR will be an indispensable part of CTR 2.0, and will take the lead in some programs, while playing an active support role in others.

NEW CHALLENGES AND OPPORTUNITIES

The committee believes that DOD can make major contributions to meeting security challenges in the Middle East, Asia, the DPRK, or other regions and countries through skillful application of its established expertise and the development of new approaches. The 2009 Annual Threat Assessment of the Intelligence Community for the Senate Select Committee on Intelligence characterizes the region from the Middle East to South Asia as an "Arc of Instability" and "the locus for many of the challenges facing the United States in the twenty-first century."[3] This assessment argues in favor of looking closely at what engagement opportunities exist or may be developed under CTR 2.0.

---

[2] U.S. Congress. Public Law 109-401. Available as of March 2009 at http://thomas.loc.gov/cgi-bin/bdquery/z?d109:HR05682:@@@L&summ2=m&|TOM:/bss/d109query.html|.

[3] Dennis C. Blair. 2009. Testimony at the 2009 Annual Threat Assessment of the Intelligence Community for the Senate Select Committee on Intelligence. 8 pp.

Each new engagement will take place in the context of broader U.S. policy, and diplomacy will likely have to lay the groundwork for any new CTR undertakings. Programs will have to be designed against a complex set of political, social, economic, and security conditions. In the Middle East, the level of tension between Israel and the Palestinians and the role that the United States plays in that conflict can affect whether or not countries in the region choose to engage with the United States. Another key element is Iran. Several recent studies[4] have concluded that Iran's nuclear program, should it proceed to a nuclear arsenal, could lead to a cascade of proliferation in the region. Should this situation develop it will provide both a challenge and an opportunity for CTR 2.0. In South Asia, the cycle of wars and crises since the 1949 partition of India has only been heightened by the November 2008 attacks in Mumbai, India. Although the environment is unstable, India and Pakistan both have nuclear weapons, chemical and biological industrial capabilities, and missile programs. They may be good candidates for engagement under CTR 2.0, but finding the right place to start and being willing to begin with modest efforts to build trust and confidence may be all that can be expected at the outset. The DPRK presents special challenges because of its authoritarian regime, lack of transparency, isolation, and a history of not living up to international commitments. But the United States and others are actively pursuing nuclear disablement and dismantlement that could lead to a broader set of CTR-type programs. In the following sections, the committee lists some challenges and opportunities for DOD CTR as a contributor to CTR 2.0.

POSSIBLE ROLES FOR DOD CTR IN SUPPORTING INTERNATIONAL AND MULTILATERAL SECURITY INITIATIVES

The DOD CTR program played an important role in helping Russia and other countries in the FSU fulfill international nonproliferation treaty obligations, particularly those related to the Strategic Arms Reduction Treaty. Once Belarus, Kazakhstan, and Ukraine decided to give up their nuclear weapons, the DOD CTR program provided substantial and critical assistance to repatriate nuclear warheads safely and securely to Russia, and worked with Belarus, Kazakhstan, and Ukraine, to eliminate WMD, delivery systems and infrastructure. It supported nuclear warhead deactivation, secure storage of the warhead fissile material, and continues its work to destroy Russia's very substantial stockpile of chemical weapons as a partner in the international program under the Group of Eight Global Partnership (G8 GP). Based on DOD CTR's successful activities in treaty implementation, it should provide similar support on

---

[4] International Institute for International Studies. 2008. *Nuclear Programs in the Middle East – In the Shadow of Iran.* 9 pp. Available as of March 2009 at http://www.iiss.org/publications/strategic-dossiers/nuclear-programmes-in-the-middle-east-in-the-shadow-of-iran/.

a global basis under CTR 2.0 both to traditional arms control treaties and to new international security mechanisms.

The committee identified several opportunities during its deliberations and in its discussions with experts that demonstrate the breadth of global engagement potential under CTR 2.0. Table 4.1 reflects possible engagement areas and suggests partners that could contribute to each activity. Some activities may require sensitive negotiations before program activities can begin, such as chemical weapons stockpile destruction projects; others may be able to begin in the very near term by expanding on existing programs, such as industrial chemical safety and security.

PROMOTING IMPLEMENTATION OF THE CHEMICAL WEAPONS CONVENTION

In the area of traditional arms control treaties, there is significant potential under the Chemical Weapons Convention (CWC) for expanded DOD CTR activity.

Chemical Weapons Destruction

The chemical weapons arsenals in the Middle East could be a prime target area for CTR 2.0. Large stockpiles of chemical weapons are believed to exist

TABLE 4.1 Examples of Possible DOD CTR 2.0 Activities

Region or Country	Activity	Countries	Possible Partners
Middle East	Secure and destroy suspected chemical weapons stockpiles	Egypt Israel Syria	DOD G8 GP Organisation for the Prohibition of Chemical Weapons (OPCW)
	Eliminate remaining chemical weapons munitions and develop technical capability to eliminate any future chemical weapons	Iraq	DOD Environmental Protection Agency (EPA) OPCW
	Promote Chemical Weapons Convention (CWC) Accession, including – chemical weapons detection and interdiction equipment and training – training for parliamentarians and national technical advisors	Lebanon Iraq	DOD Department of State G8 GP OPCW

TABLE 4.1 Continued

Region or Country	Activity	Countries	Possible Partners
Middle East, Africa, Asia	Promote industrial chemical safety and security – protecting chemical facilities – protecting cargoes of hazardous chemicals in transit	Multiple countries	DOD Department of State Department of Homeland Security G8 GP OPCW Industry Nongovernment Organizations (NGOs)
	Promote biological safety, security, and disease surveillance programs	Multiple countries	DOD Department of State Department of Health and Human Services (HHS) G8 GP World Health Organization (WHO) Industry
	Promote United Nations Security Council Resolution 1540 implementation – countersmuggling, counterpiracy, countertrafficking	Multiple countries	DOD Department of State DOE G8 GP Countries involved in Proliferation Security Initiative and Global Initiative to Combat Nuclear Terrorism
	Promote Defense and Military Contacts (DMC) programs in more nations; connect DMC State Partnership Programs with CTR-related activities	Multiple countries	DOD DMC Countries involved
	Facilitate incident/emergency response training programs	Multiple countries	DOD DOE Department of State HHS Countries involved
	Develop cybersecurity training programs	Multiple countries	DOD DOE Countries involved
	Strengthen export controls and border security, including maritime security	Multiple countries	DOD Department of State DOE Coast Guard Countries involved

TABLE 4.1 Continued

Region or Country	Activity	Countries	Possible Partners
Asia, Africa, Central Europe	Secure and eliminate excess conventional munitions	Philippines, multiple countries in Africa, Albania	DOD Department of State
DPRK	Provide environmental monitoring laboratory equipment and training for nuclear contamination assessment and decontamination at Yongbyon (including redirection of former weapons scientists)		DOD DOE Department of State Russia, South Korea, Japan, China International Atomic Energy Agency (IAEA)
	Provide logistical support for denuclearization		DOD DOE Department of State IAEA
Asia	Promote biological safety, security, and disease surveillance programs	Pakistan Indonesia	DOD Department of State HHS G8 GP WHO Industry NGOs
	Promote chemical safety and security – protecting chemical facilities – protecting cargoes of hazardous chemicals in transit	India Pakistan	DOD Department of State EPA OPCW Industry NGOs
	Facilitate incident/emergency response planning and training	Multiple countries	DOD DHS DOE
	Facilitate scientist-to-scientist exchanges in support of nonproliferation technologies	India Pakistan	Department of State DOE DOD Academic community
Russia	Complete all current projects with emphasis on sustainability		DOD DOE
	Coidentify lessons learned and best practices as basis of a strategy for application of programs outside the FSU		DOD Department of State DOE G8 GP Counterpart Russian agencies

in Egypt, Israel, and Syria, none of which have joined the CWC. (Israel signed, but did not ratify the convention.) Bringing these three states, along with Iraq and Lebanon, under the disciplines entailed in CWC membership could help reduce tensions in the region.

Despite pledges by Egypt, Israel, and Syria to work toward elimination of WMD in the region,[5] it is unlikely that these states will give up their chemical weapons without pressure or incentives from some outside party such as the United Nations, the United States, or the nascent Mediterranean Union (or, in Israel's case, absent a regionwide resolution of the Middle East conflict). Any initial U.S. government actions with these countries would seem most appropriate for the State Department, perhaps as an initiative under the G8 GP, which has played a strong role in providing assistance to Russia's chemical weapons destruction effort. DOD CTR might play a role later by providing technical expertise in the destruction of chemical weapons. As part of the chemical weapons destruction process, DOD CTR could establish or strengthen environmental monitoring capabilities for toxic chemicals that could be left in place once the chemical weapons destruction is finished. This could also be supported by the Environmental Protection Agency. DOD CTR could also provide emergency response training and personal protective equipment for units in partner countries that would be called upon to respond to chemical exposure, whether intentional or accidental.

Some 500 chemical weapons munitions escaped destruction in Iraq during the UN-supervised campaign after the first Gulf War.[6] These weapons need to be destroyed as part of Iraq's responsibilities under the CWC, as well as to keep them from falling into the hands of terrorist organizations. DOD, through its Chemical Materials Agency and contractors, can provide extensive expertise and assistance in destroying these weapons, if requested by Iraq's National Authority. DOD CTR can also train Iraqi units in destruction techniques and leave behind a permanent capability in Iraq that can manage any future discovery of additional chemical weapons stockpiles. Such units, when properly trained and with experience working on their own chemical weapons destruction, could also offer similar assistance to other countries in the region. Middle East states may accept assistance from Iraqi experts more readily than from the U.S. or European sources.

Promoting Accession to the CWC

The DOD CTR program can play different but important roles in promoting the accession of Iraq and Lebanon to the CWC, both of which have taken

---

[5] Summit of Mediterranean States, Paris, July 13, 2008.
[6] National Ground Intelligence Center (NGIC). 2006. Unclassified excerpt from NGIC, released to House Intelligence Committee, June 21.

steps toward joining.[7] Program focus would be on providing technical expertise in the detection, handling, and destruction of chemical weapons. In Iraq, the primary role for DOD CTR would be to provide assistance in the destruction of legacy chemical weapons from Saddam Hussein's regime and related training of Iraqi units.

The Organization for the Prohibition of Chemical Weapons (OPCW) can provide training for parliamentarians and technical experts who will be responsible for the ratification of the convention and enacting the necessary implementing legislation. The latter legislation may be critical for Lebanon to prevent the transport of chemical weapons onto its territory.

Lebanon, unlike Iraq, has no known chemical weapons at present. It could, however, become a threat to security in the region if Hezbollah forces in southern Lebanon were to acquire chemical warheads for short-range missiles such as the ones that it fired into Israel in the short-lived 2006 war.[8] Presumably, the CWC implementing legislation will be structured to prohibit moving chemical weapons onto Lebanese territory. However, enforcement of such prohibitions will require technical expertise and equipment to detect and interdict transfers of chemical weapons material from a neighboring state such as Syria or Iran. In this case, the DOD CTR Proliferation Prevention Initiative, working with other U.S. CTR programs involved in border security, should be able to provide expertise and equipment if requested by Lebanon's National Authority or the United Nations Interim Force in Lebanon force in the area.

Overall, DOD CTR should be able to provide significant technical assistance to any nation seeking to join the CWC. The nature of the cooperation will have to be tailored to the needs and sensitivities of each state. Close cooperation with the OPCW will be essential; coordination with the G8 GP will also be necessary and may also lead to opportunities to share the costs of program implementation. The broad partnership that has supported chemical weapons destruction in Russia will ensure broader international commitment of technical and financial resources.

Reducing the Risk of Chemical Attack

Several experts commented to the committee that in their view the risk of chemical attack is underestimated and consequently receives far too little attention. DOD CTR could make significant contributions to the stability of

---

[7] Global Security Newswire. 2008. Lebanon Joins Chemical Weapons Convention. Washington, D.C.: Nuclear Threat Initiative. December 1. Available as of March 2009 at http://www.globalsecuritynewswire.org/gsn/nw_20081201_8457.php.

[8] Nissan Ratzlav-Katz and Pinchas Sanderson. 2008. Hizbullah Gears Up for War, Olmert Asks for UN Help. *Arutz Sheva*. July 14. Available as of March 2009 at www.Israelinternationalnews.com/news/news.aspx/126842.

countries facing domestic terrorism by preventing terrorist acquisition of toxic industrial chemicals as shown in the examples below.

Potential *releases of toxic gases* from chemical plants or refineries pose significant risks to the populations of many cities, particularly in the Middle East, Africa, and parts of Asia where chemical industries have developed, but without the benefit of rigorous industrial safety standards. Deliberate releases through sabotage, terrorist, or militant attacks could threaten the stability of many nations, which in turn could have a direct impact on U.S. security. The DOD CTR program has technical expertise and experience to help counter such threats in other countries.

Protecting chemical plants or refineries is an area in which the DOD CTR program can draw on its experience acquired in safeguarding nuclear, chemical, and biological facilities in the FSU. Similarly, DOE could contribute by drawing on its relevant experience in protecting nuclear facilities in the United States and internationally. The potential problems are similar to those confronting the Department of Homeland Security (DHS) in its assessment of the vulnerability of industrial facilities in the United States. DHS, along with the State Department, which has already engaged many international partners on the chemical security issue, can be partners in such an effort.

Protecting cargoes of hazardous chemicals in transit and storage is especially challenging when the materials are being moved on public highways or through ports handling large volumes of commercial cargo. Again, the DOD CTR program should be able to draw on its experience in the FSU to assist partner states in developing technology to safeguard chemical shipments. Likewise, DOE has extensive experience in safeguarding the domestic transportation of nuclear materials. The protection of hazardous materials in storage would bring special challenges for working with commercial, in addition to governmental, facilities. It would require a high degree of flexibility in developing genuine partnership arrangements. Such an effort could be assisted by chemical industry trade associations that have developed best practices for the handling of dangerous materials.

In all these areas, DOD CTR could add its special expertise in the chemical security area to that of the State Department, which launched a Chemical Security Engagement Program in 2007,[9] as a companion program to its Biological Security Engagement Program.[10] This new effort implements programs in conjunction with host governments to fill critical gaps in chemical security and safety, particularly where there is high potential for terrorist activity.

In August 2007, the State Department teamed with the International Union

---

[9] Chemical Security Engagement Program. Available as of March 2009 at http://www.csp-state.net/dev/contact/index.aspx.

[10] Department of State. Biosecurity Engagement Program. Available as of March 2009 at http://ironside.sandia.gov/AsiaConference/JasonRao-BEP.pdf.

for Pure and Applied Chemistry (IUPAC), an international scientific association, to organize a 1-day workshop in Kuala Lumpur, "Chemical Safety and Security in the 21st Century." The objective of the workshop was as follows: "To raise awareness of the chemical threat and to identify gaps in chemical security and chemical safety practices in South and Southeast Asia among practicing chemists, governmental officials, and regional chemical industry representatives."[11] IUPAC was selected as a partner for the effort because of its track record of working with the OPCW on similar issues and its international network.

Information gathered from this and similar conferences could provide the basis on which the DOD CTR program could explore developing initiatives in this area. This could be a fertile ground for new efforts that build on the extensive experience DOD CTR has had in chemical weapons destruction and security. Under CTR 2.0, similar collaborations with other international scientific unions and organizations could also be explored. For example, the U.S. National Academy of Sciences, which facilitated the link between the State Department and IUPAC, also is a member of the International Council for Science (ICSU) and oversees a network of more than 20 U.S. national committees corresponding to various ICSU scientific member bodies.[12]

IMPLEMENTING UNSCR 1540

The passage of United Nations Security Council Resolution 1540 (UNSCR) and the reinforcement of those principles in UNSCR 1810 provide a new range of potential DOD CTR activity. For example, UNSCR 1540 addresses the issue of chemical weapons proliferation to nonstate organizations much more directly than does the CWC Article VII, which was negotiated with state players in mind. Both documents dictate implementing legislation that prohibits persons or parties within territory under the control of the member state from possessing or producing chemical weapons. The UNSCR goes beyond the CWC in many ways, particularly because it is binding on all states, not just the signatories of the CWC. One clause is particularly relevant to the current discussion.

---

[11] International Union of Pure and Applied Chemistry (IUPAC). 2007. Project: Chemical Safety and Security in the 21st Century. Available as of March 2009 at http://www.iupac.org/web/ins/2007-021-2-020. As stated on the IUPAC Web site, the workshop goals were to "1. Gain understanding about gaps in chemical security and chemical safety as identified by Governmental officials, practicing chemists, industry representatives, and international experts, with a particular focus on South and Southeast Asia; 2. Investigate ways in which IUPAC, other international organizations, and the State Department Chemical Security Engagement Program could develop programming to work with host governments, practicing chemists, local and regional chemical organizations, and chemical industry to begin to fill gaps. Follow on efforts could include best practices training, risk management strategy sharing, and cooperative research and development; and 3. Raise awareness of chemical terrorism threat among practicing chemists and industry in South and Southeast Asia."

[12] For more information, see the National Academies Board on International Scientific Organizations as of March 2009 at www.nas.edu/biso.

(The UNSCR) recognizes that some States may require assistance in implementing the provisions of this resolution within their territories and invites States in a position to do so to offer assistance as appropriate in response to specific requests to the States lacking the legal and regulatory infrastructure, implementation experience and/or resources for fulfilling the above provisions.[13]

In situations like that described for Lebanon, CTR 2.0 has a clear opportunity to assist the local Lebanese National Authority to carry out its responsibilities under UNSCR 1540. DOD CTR can work with the OPCW to provide the technical training, advice, and equipment resources needed to perform the monitoring and interdiction functions of the National Authority.

DOD CTR efforts to increase biological safety and security can also be expanded. The Department of State, working with the Departments of Health and Human Services and Agriculture, has already started biological safety and security activities in many countries in Asia and Africa. DOD CTR could explore how it can contribute to strengthening and expanding those programs and especially how it can employ its expertise in the biosecurity area that other USG CTR programs lack.

The Departments of Defense and Energy have shared responsibility for nuclear security issues in the FSU for many years. To some extent, there is a tacit division of labor, with DOE responsible for programs that address civil nuclear materials and related issues and DOD responsible for the military side. The responsibilities are not precisely defined, but the two departments appear to be able to divide the work without major dispute or duplication. As DOD and DOE look toward new activities coordinated under CTR 2.0, it may be possible for both to take a role in emerging areas of concern, for example, in limiting nuclear weapons proliferation that may result from the global expansion of nuclear power.

The projects above illustrate that new DOD CTR opportunities will likely be smaller and more varied than CTR 1.0 projects. Chemical security projects in Pakistan may differ widely from those in the Philippines, and each will have to be designed to fit local needs and capabilities.[14] The projects will also require the skills of other entities, including other U.S. government agencies, multinational organizations, and even nongovernment organizations. Effectiveness will require thoughtful integration of U.S. and international partners into each project. To bring these partners together as teams will require the hallmark CTR 2.0 characteristics of nimbleness and flexibility.

---

[13] A technical advisor to the UNSCR 1540 Committee confirmed that there currently is no mechanism for responding to technical assistance requests. Although the committee is informed of several activities that support UNSCR 1540 implementation, there is no systematic way of documenting these activities or the countries that are providing or receiving technical support.

[14] Carson Kuo, State Department, and Nancy Jackson, DOE. 2008. Communication to Committee, October 15.

One of the key lessons learned from the experiences of chemical weapons destruction in Albania and Libya is that DOD CTR and the State Department's Nonproliferation and Disarmament Fund must work together as a team. This will be even more the case under CTR 2.0. Each program developed specific skills and capabilities that complement those of the other. Together they can move a CTR project forward faster, more smoothly, and more cost-effectively than when acting independently.

DOD CTR also has important potential partners in other DOD programs. For example, the Defense Threat Reduction Agency (DTRA) has programs in International Counterproliferation, Counternarcotics, Consequence Management, Nuclear Forensics, and Small Arms and Light Weapons. In the broader CTR 2.0 environment, DOD CTR can draw on all of these as program partners. Some of these programs already are active in Unified Combatant Commands, although in a very limited way. In addition, the Department of Defense has about 20 programs that deal with some aspect of health. It would be useful to look at each to see if there is the potential for partnership with DOD CTR.

Even though WMD and their related materials, technologies, expertise, and delivery systems will always be a priority, there are many other threats that CTR 2.0 must address. As early as 1993, Congress recognized that destabilizing conventional weapons should also be covered under DOD CTR. The importance of this threat was highlighted again by the collaboration of Senators Richard Lugar and Barack Obama to pass the Department of State Authorities Act of 2006, under which Section 11 authorizes the secretary of state to secure, remove, or eliminate stocks of conventional weapons.[15] Applying security and destruction programs to unguarded stockpiles of conventional munitions may help prevent terrorist acquisition of the raw materials needed for improvised explosive devices, which have taken far more lives in Iraq than any WMD and could appear anywhere else the materials and know-how is available. CTR 2.0 can provide the opportunity for DOD CTR and State Department programs to work together in this area.

Finding 4-2: Full integration of DOD into CTR 2.0, working in concert with other U.S. government departments and within DOD, will enable DOD to make a more effective contribution to U.S. threat reduction efforts.

Recommendation 4-1: As CTR 2.0 engagement opportunities emerge, the White House should determine the agencies and partners that are best suited to execute them, whether by virtue of expertise, implementation capacity, or funding.

---

[15] Public Law 109-472 strengthens U.S. efforts to interdict illicit shipments of weapons or materials of mass destruction and secure vulnerable stockpiles of conventional weapons.

OPTIMIZING THE DEFENSE AND MILITARY CONTACTS PROGRAM

The Defense and Military Contacts (DMC) program funded under CTR 1.0 was not used historically to advance the DOD CTR program. Although funded by the DOD CTR budget, the DMC program was initially directed by the office of the Deputy Assistant Secretary of Defense for Eurasia, Russia, and Ukraine, which reported to the office of the Assistant Secretary of Defense for International Security Policy, with little involvement of the DOD CTR policy office. DMC is well suited to supporting engagements with new partners under CTR 2.0. It includes several activities that could be expanded to have broader application and that also respond to priorities of both Unified Combatant Command and DOD CTR missions. Officers at several Unified Commands expressed a high degree of interest in the following types of existing DMC activities:

- Traveling Contact Teams (TCTs) for maritime interdiction and nuclear, biological and chemical warning and detection
- Military Police familiarization exchanges and antiterror TCTs
- National Guard State Partnership Program familiarizations and contact visits
- Regional counterproliferation and counterterrorism exercises
- Disaster preparedness and consequence management TCTs

The DMC program could be administered directly as part of the overall DOD CTR program and be used to lay the groundwork for future CTR 2.0 engagements. Future DMC program planning would benefit from direct engagement with the Unified Commands, within an overall strategic framework and in close coordination with diplomatic and other efforts.

Finding 4-3: The Defense and Military Contacts Program, funded by DOD CTR, is a relatively small, but potentially important, element of the DOD CTR 2.0 effort and could be better focused to support specific DOD CTR relationship-building opportunities that lead to program development in new geographic areas.

A ROLE FOR THE UNIFIED COMBATANT COMMANDS

The Unified Combatant Commands,[16] particularly those with geographic responsibility, are well positioned to help identify potential CTR 2.0 activities.

---

[16] The Unified Commands include U.S. Northern Command, U.S. Pacific Command, U.S. Southern Command, U.S. Central Command, U.S. European Command, U.S. Joint Forces Command,

Because the focus of CTR 1.0 was on Russia and the FSU, the commands, other than the European Command, have not been involved in CTR programs, are not part of the planning process, and even are unaware of many CTR 1.0 activities in their areas of responsibility. The commands, however, already have aspects of CTR 2.0 in their operations plans and even in some projects they support. For example, U.S. Pacific Command (PACOM) participates in a joint avian influenza surveillance project with the U.S. Agency for International Development as part of its biodefense effort and had sponsored a biological security workshop in Malaysia. At the time of the committee's conversation at PACOM, officers there were unaware of the DOD CTR Biological Threat Reduction Program or that the Department of State was engaged in biosecurity activities in the Pacific area.

As the newest regional command, the U.S. African Command (AFRICOM) presents a particularly interesting opportunity to build on existing relationships that might be a model for other commands. AFRICOM faces the daunting task of balancing demands to prevent global terrorism from taking hold in an environment of poverty, poor education, massive population growth, and health challenges. Many African nations and international organizations are reluctant to encourage further militarization of the continent. Africa holds a significant portion of the world's natural resources, including vast untapped reservoirs of oil, making it a focal point of global interests as energy demands rise, driven especially by countries with rapidly increasing standards of living such as China and India.

> DOD could build on its long-standing presence in Africa established by the medical research units of the U.S. Navy in Cairo, Egypt, and the U.S. Army in Nairobi, Kenya.

How the "face" of AFRICOM is developed now will influence how successful it will be in the years to come. Fortunately, DOD has a long-standing presence in Africa established in large part by the medical research units of the U.S. Navy in Cairo, Egypt, and the U.S. Army in Nairobi, Kenya. Both of these programs have been in place for decades, have built solid foundations of collaboration and mutual respect between their respective organizations and their host governments, and in many cases these facilities have served as launch sites for outreach activities and outbreak investigation into other countries in Africa and beyond, including into the central Asian states of the former Soviet Union.

The focus of activities at the medical research laboratories has been on

---

U.S. Special Operations Command, U.S. Transportation Command, and U.S. Strategic Command. The Unified Commands are referred to collectively in the report as "Commands."

endemic diseases that are also of military concern, such as malaria, leishmaniasis, and others, and have extended on occasion to work on pathogens with bioterrorism potential, such as anthrax, plague, and Ebola. Over the years, strong partnerships have developed between U.S. military and civilian scientists and physicians with their local collaborators, resulting in shared authorship of scientific publications and the establishment of life-long friendships. As measured by most metrics of productivity, transparency, and engagement, these laboratories have been highly successful.

Some of the diseases studied at these facilities are now the focus of large global initiatives that involve important NGOs such as the Bill and Melinda Gates Foundation and the Global Fund to Fight AIDS, Tuberculosis and Malaria.[17] The DOD overseas laboratories have also worked closely with the World Health Organization in responding to outbreaks of global importance that have occurred in the region, and in other international collaborations such as the global surveillance of seasonal and avian influenza. The DOD overseas medical research laboratories are also closely linked to the U.S. Centers for Disease Control and Prevention's (CDC) international activities, with CDC staff members frequently assigned to the DOD overseas laboratories. In Kenya, the CDC has staffed its own medical research laboratory with its headquarters in Nairobi and a robust field site in Kisumu in western Kenya. The CDC laboratory has been present in Kenya for about three decades and is very well regarded. Other smaller profile activities under CDC's direction are in place in Tanzania, Uganda, and until recently in Cote d'Ivoire, and CDC staff members can be found in many African nations assisting with childhood immunization programs, the global eradication of polio campaign, and other global health initiatives. Collectively, these activities present the United States in a very positive light locally, and could offer AFRICOM a foundation to build upon that could both help address important global health challenges and provide access to critical local information and early warning of disease problems. An important challenge to AFRICOM will be to make certain that their mission of terrorism prevention does not negate the longstanding good will established by these highly successful resident programs.

The committee consulted with several commands to explore how aware they are of existing DOD CTR and U.S. government CTR efforts and the extent to which a CTR 2.0 might be integrated into command strategies. The level of interest was high, as was the potential relevance of CTR 2.0 to command missions. DTRA currently has liaison officers stationed at each of the commands who could provide a ready link between DOD CTR and command interests. In addition to keeping commands informed of DOD CTR programs, these liaisons, if incorporated into the broader flow of information from all U.S. government participants in CTR 2.0, could share that information as well.

---

[17] See http://www.theglobalfund.org/en/.

Finding 4-4: Combatant commands currently engage regionally at many levels and with a broad group of interlocutors, but too little with DOD CTR or other U.S. government departments implementing cooperative threat reduction programs. DOD-DTRA cultivation of relationships with the combatant commands creates mutual benefits.

Recommendation 4-2: The secretary of defense should direct the review and reformulation of the DOD CTR program in support of CTR 2.0 and work with the White House, secretary of state, secretary of energy, and other cabinet and agency officers to ensure full coordination and effective implementation of DOD programs in CTR 2.0. The review should also include broader military components, including the Unified Combatant Commands, the full set of programs in DTRA, DOD health and research programs, and other DOD assets.

The substantial changes in form and function proposed for CTR 2.0 will not be implemented overnight. Many existing program commitments must be fulfilled, and fundamental changes in how U.S. government agencies relate to each other and how the U.S. government relates to its domestic and international partners will take time. Many lessons have been learned from the CTR 1.0 experience in the FSU that need to be remembered; best practices need to be applied while new ones are developed. The White House, working across agencies and with Congress, needs to devise a plan that will allow CTR 2.0 to be constructed while the United States completes its commitments under CTR 1.0. There may be some programs in CTR 1.0 that can evolve earlier than others, and these should be encouraged. There may be examples from the G8 GP that can be held up as an example to other nations as a model. The key is to begin the process and not wait for the next emergency, then wish that CTR 2.0 was there to respond.

Recommendation 4-3: A plan for the evolution of CTR 1.0 to CTR 2.0 should take into account the congressional principles enumerated in the legislation authorizing this report, as well as existing USG CTR initiatives. The White House should review National Security Council–Homeland Security Council coordination in bioengagement as a possible model for other programs as it develops a transition plan.

SUMMARY OF CHAPTER FINDINGS AND RECOMMENDATIONS

Finding 4-1: DOD CTR will be an indispensable part of CTR 2.0, and will take the lead in some programs, while playing an active support role in others.

Finding 4-2: Full integration of DOD into CTR 2.0, working in concert with other U.S. government departments and within DOD, will enable DOD to make a more effective contribution to U.S. threat reduction efforts.

Recommendation 4-1: As CTR 2.0 engagement opportunities emerge, the White House should determine the agencies and partners that are best suited to execute them, whether by virtue of expertise, implementation capacity, or funding.

Finding 4-3: The Defense and Military Contacts Program funded by DOD CTR, is a relatively small, but potentially important, element of the DOD CTR 2.0 effort and could be better focused to support specific DOD CTR relationship-building opportunities that lead to program development in new geographic areas.

Finding 4-4: Combatant commands currently engage regionally at many levels and with a broad group of interlocutors, but too little with DOD CTR or other U.S. government departments implementing cooperative threat reduction programs. DOD-DTRA cultivation of relationships with the combatant commands creates mutual benefits.

Recommendation 4-2: The secretary of defense should direct the review and reformulation of the DOD CTR program in support of CTR 2.0 and work with the White House, secretary of state, secretary of energy, and other cabinet and agency officers to ensure full coordination and effective implementation of DOD programs in CTR 2.0. The review should also include broader military components, including the Unified Combatant Commands, the full set of programs in DTRA, DOD health and research programs, and other DOD assets.

Recommendation 4-3: A plan for the evolution of CTR 1.0 to CTR 2.0 should take into account the congressional principles enumerated in the legislation authorizing this report, as well as existing USG CTR initiatives. The White House should review National Security Council–Homeland Security Council coordination in bioengagement as a possible model for other programs as it develops a transition plan.

5

Cooperative Threat Reduction 2.0: Implementation Checklist

The congressional request for this report expressed the Sense of Congress that the Department of Defense (DOD) Cooperative Threat Reduction (CTR) program should be strengthened and expanded and that part of this process should be the development of new initiatives. The committee believes that its vision of CTR 2.0 is achievable and is fully consistent with Congress's view. This report's findings and recommendations identify actions that are necessary to begin the transition process toward CTR 2.0. The White House, working in concert with senior officials in the administration and Congress, should initiate the actions in the implementation checklist below:

- Establish who in the White House will lead the CTR 2.0 effort and solicit active support of senior officials in the relevant departments and agencies.
- Develop a clear strategy and country-region priorities, taking into account resources available across the government and through nongovernment and international partners. The strategy should include the following:

    o rapid response capability as well as the capacity to implement longer-term programmatic efforts

    o specific ways that CTR 2.0 can support both traditional arms control and nonproliferation treaties as well as new multilateral and international security instruments

    o involvement of a wide range of government, nongovernment, and international partners from the outset

    o consideration of other strategies for the same country or region, for example, from Unified Combatant Commands, the U.S. Agency for International Development, and United Nations organizations

- Work with the Office of Management and Budget to do the following:

    o Develop meaningful program metrics that

    - highlight program impact
    - acknowledge the value to national security of intangible program results
    - incorporate partner metrics into the overall evaluation of programs
    - link metrics to program selection criteria

    o Establish a multiyear budget planning basis for CTR 2.0 programs and develop a process for allocating budgets to the agencies needed to implement programs

- Encourage a new generation of congressional leaders to engage actively through regular consultations between the executive and legislative branches and work with Congress to do the following:

    o add authorities to the departments whose participation is essential to CTR 2.0, but that are considered nonsecurity agencies

    o allocate funding for security programs directly to the agencies that will be responsible for implementation

    o provide the ability to all agencies implementing CTR 2.0 to comingle other funds with congressionally appropriated funds

    o provide "notwithstanding authority" to up to 10 percent of the DOD CTR annual appropriation to ensure the capacity to respond rapidly and flexibly to opportunities

- Identify ways to make programs less cumbersome and bureaucratic, and more timely, agile, and responsive to partner priorities.

    o review the DOD CTR umbrella agreement and other U.S. government contracting mechanisms to assess what is required and where flexibility can be introduced

    o review current USG CTR programs to see which ones already implement elements of CTR 2.0 and can be used as models

- Continue working with established partners and identify new partners–develop a "habit of cooperation."

    o demonstrate partnership by beginning new program engagements with information sharing, joint identification of risks and opportunities, collaborative planning, and shared responsibilities for project leadership, management, and resources

    o use the habit of cooperation as the basis for program transparency

- Promote sustainability by doing the following:

    o incorporating sustainability into the first stages of program development
    o engaging partners in program development, design, and implementation
    o seeking the input of local officials, scientists, and nongovernment organizations to ensure that programs are relevant and will sustain the interest and commitment of the local partners

The legislative language also specifies eight elements listed below that new initiatives should include as they are developed and implemented. The committee considered these elements, most of which are addressed specifically in this report and all of which apply broadly to CTR 2.0. This list provides a useful point of departure for any guidelines that the administration may prepare in the future and reflects important insights and lessons learned from the nearly two decades of program experience. These points are also listed in Appendix J along with the findings and recommendations relevant to them.

1. Programs should be well coordinated with the Department of Energy (DOE), the Department of State, and any other relevant U.S. government agency or department.
2. Programs will include appropriate transparency and accountability mechanisms, and legal frameworks and agreements between the United States and CTR partner countries.
3. Programs should reflect engagement with nongovernmental experts on possible new options for the CTR program.
4. Programs should include work with the Russian Federation and other countries to establish strong CTR partnerships. Among other things, these partnerships should

    (i) Increase the role of scientists and government officials of CTR partner countries in designing CTR programs and projects; and
    (ii) Increase financial contributions and additional commitments to CTR programs and projects from Russia and other partner countries, as appropriate, as evidence that the programs and projects reflect national priorities and will be sustainable.

5. Programs should include broader international cooperation and partnerships, and increased international contributions.

6. Programs should incorporate a strong focus on national programs and sustainability, which includes actions to address concerns raised and recommendations made by the Government Accountability Office, in its report of February 2007 titled "Progress Made in Improving Security at Russian Nuclear Sites, but the Long-Term Sustainability of U.S. Funded Security Upgrades is Uncertain," which pertain to the Department of Defense.

7. Efforts should continue to focus on the development of CTR programs and projects that secure nuclear weapons; secure and eliminate chemical and biological weapons and weapons-related materials; and eliminate nuclear, chemical, and biological weapons-related delivery vehicles and infrastructure at the source.

8. There should be efforts to develop new CTR programs and projects in Russia and the former Soviet Union, and in countries and regions outside the former Soviet Union, as appropriate and in the interest of U.S. national security.

DEVELOPING A NEW GENERATION OF GLOBAL SECURITY ENGAGEMENT EXPERTS

The committee was not asked to consider the issue of staffing, but believes that it is important to the overall discussion of future programs.[1] The initial set of government officials, scientists, and engineers who created CTR 1.0 have mentored new staff over time, but the nature of CTR programs is highly technical, and in general, the academic specialties of science and security have been poorly integrated into program staffs. While the skills that went into creating CTR 1.0 are highly valued, CTR 2.0 will demand some new capabilities. CTR 1.0 program models have been developed and tested, implementation approaches have succeeded and failed, best practices have been identified, and many lessons learned can save both time and resources in future engagements. At a minimum, consideration might be given to developing a course, perhaps organized jointly by the Departments of Defense, State, and Energy and other implementing departments and offered through the Department of State Foreign Service Institute or through the National War College or both. Specifically for DOD, layers of courses could be developed that would include command and staff colleges, service war colleges, and a National Defense University Capstone course. DOE could also incorporate specific training into its courses. One of the aims of such training experiences would be to bring

---

[1] This issue is also addressed by the Project on National Security Reform. See Project on National Security Reform. 2008. Forging a New Shield. Availabe as of March 2009 at http://www.pnsr.org/data/files/pnsr%20forging%20a%20new%20shield.pdf.

individuals together from across departments with the goal of team building and better understanding each other's missions, capabilities, and programs as participants in CTR 2.0. Team building can be further amplified by providing regular opportunities for staff to be detailed to other agencies' CTR programs. Providing training and interagency exchange opportunities that help define a career path may also help attract new talent to the CTR area. As it is, the policy offices at the Departments of Defense, State, and Health and Human Services and other agencies are staffed by dedicated professionals, but there are too few professionals to do the job that is required under CTR 2.0.

List of Acronyms

AFRICOM U.S. Africa Command
ASM Air-to-Surface Missiles

BS&S Biosecurity and Biosafety
BTRP Biological Threat Reduction Program

CBR Cooperative Biological Research
CDC Centers for Disease Control and Prevention
CTR Cooperative Threat Reduction
CWC Chemical Weapons Convention
CWD Chemical Weapons Destruction

DASD/ISP Deputy Assistant Secretary for Defense for International Security Policy
DMC Defense and Military Contacts Program
DOD Department of Defense
DOD CTR Department of Defense Cooperative Threat Reduction
DOE Department of Energy
DPRK Democratic People's Republic of Korea
DTRA Defense Threat Reduction Agency

EDP Especially Dangerous Pathogens
EPA Environmental Protection Agency
EU European Union

FMSF Fissile Material Storage Facility
FSU Former Soviet Union

FTE	Full-Time Equivalent
G8	Group of Eight
G8 GP	G8 Global Partnership
GAO	Government Accountability Office
GICNT	Global Initiative to Combat Nuclear Terrorism
GP	Global Partnership
GSE	Global Security Engagement
HEU	Highly Enriched Uranium
HHS	Department of Health and Human Services
HSC	Homeland Security Council
IED	Improvised Explosive Device
IAEA	International Atomic Energy Agency
ICBM	Intercontinental Ballistic Missiles
ICP	International Counterproliferation
ISTC	International Science and Technology Center
IUPAC	International Union for Pure and Applied Chemistry
JVE	Joint Verification Experiment
LEU	Low-Enriched Uranium
MPC&A	Material Protection, Control and Accounting
NAS	National Academy of Sciences
NCID	National Center for Infectious Diseases
NDF	Nonproliferation and Disarmament Fund
NGO	Nongovernment Organization
NIS	Newly Independent States
NRC	National Research Council
NSC	National Security Council
NTI	Nuclear Threat Initiative
NWSS	Nuclear Weapons Storage Security Program
NWTS	Nuclear Weapons Transportation Security Program
OMB	Office of Management and Budget
OPCW	Organization for the Prohibition of Chemical Weapons
OSAC	Overseas Security Advisory Council
OTA	Congressional Office of Technology Assessment
PART	Program Assessment Rating Tool
PNSR	Project on National Security Reform

PPI	Proliferation Prevention Initiative
PSI	Proliferation Security Initiative
RDD	Radiological Dispersion Device, or Dirty Bomb
RMTC	Russian Methodological and Training Center
SAIC	Science Applications International Corporation
SLBM	Submarine-Launched Ballistic Missile
SNA	Social Network Analysis
SNAE	Strategic Nuclear Arms Elimination
SNF	Spent Nuclear Fuel
SOAE	Strategic Offensive Arms Elimination Program
SSBN	Strategic Nuclear-Powered Ballistic Missile Submarine
SSD	Safety, Security, and Dismantlement
START	Strategic Arms Reduction Treaty
STCU	Science and Technology Center in Ukraine
STC	Science and Technology Centers
TADR	Threat Agent Detection and Response
TCTs	Traveling Contact Teams
UN	United Nations
UNIFIL	United National Interim Force in Lebanon
UNSCR	United Nations Security Council Resolution
USAID	United States Agency for International Development
USAMRIID	U.S. Army Medical Research Institute for Infectious Diseases
USDA	United States Department of Agriculture
USG	United States Government
USSR	Union of Soviet Socialist Republics
WMD	Weapons of Mass Destruction
WMDIE	Weapons of Mass Destruction Infrastructure Elimination Program

Appendixes

Appendix A

H.R. 1585: National Defense Authorization Act for Fiscal Year 2008

TITLE XIII—COOPERATIVE THREAT REDUCTION WITH STATES OF THE FORMER SOVIET UNION

SEC. 1306. NEW INITIATIVES FOR THE COOPERATIVE THREAT REDUCTION PROGRAM.

(a) Sense of Congress- It is the sense of Congress that—

(1) the Department of Defense Cooperative Threat Reduction (CTR) program should be strengthened and expanded, in part by developing new CTR initiatives;

(2) such new initiatives should—

(A) be well-coordinated with the Department of Energy, the Department of State, and any other relevant United States Government agency or department;

(B) include appropriate transparency and accountability mechanisms, and legal frameworks and agreements between the United States and CTR partner countries;

(C) reflect engagement with non-governmental experts on possible new options for the CTR program;

(D) include work with the Russian Federation and other countries to establish strong CTR partnerships that, among other things—

(i) increase the role of scientists and government officials of CTR partner countries in designing CTR programs and projects; and
(ii) increase financial contributions and additional commitments to CTR programs and projects from Russia and other partner countries, as appropriate, as evidence that the programs and projects reflect national priorities and will be sustainable;

(E) include broader international cooperation and partnerships, and increased international contributions;

(F) incorporate a strong focus on national programs and sustainability, which includes actions to address concerns raised and recommendations made by the Government Accountability Office, in its report of February 2007 titled "Progress Made in Improving Security at Russian Nuclear Sites, but the Long-Term Sustainability of U.S. Funded Security Upgrades is Uncertain," which pertain to the Department of Defense;

(G) continue to focus on the development of CTR programs and projects that secure nuclear weapons; secure and eliminate chemical and biological weapons and weapons-related materials; and eliminate nuclear, chemical, and biological weapons-related delivery vehicles and infrastructure at the source; and

(H) include efforts to develop new CTR programs and projects in Russia and the former Soviet Union, and in countries and regions outside the former Soviet Union, as appropriate and in the interest of United States national security; and

(3) such new initiatives could include—

(A) programs and projects in Asia and the Middle East; and

(B) activities relating to the denuclearization of the Democratic People's Republic of Korea.

(b) National Academy of Sciences Study-

(1) STUDY- Not later than 60 days after the date of the enactment of this Act, the Secretary of Defense shall enter into an arrangement with the National Academy of Sciences under which the Academy shall carry out a study to analyze options for strengthening and expanding the CTR Program.

(2) MATTERS TO BE INCLUDED IN STUDY- The Secretary shall provide for the study under paragraph (1) to include—

(A) an assessment of new CTR initiatives described in subsection (a); and

(B) an identification of options and recommendations for strengthening and expanding the CTR Program.

(3) SUBMISSION OF NATIONAL ACADEMY OF SCIENCES REPORT- The National Academy of Sciences shall submit to Congress a report on the study under this subsection at the same time that such report is submitted to the Secretary of Defense pursuant to subsection (c).

(c) Secretary of Defense Report-

(1) IN GENERAL- Not later than 90 days after receipt of the report under subsection (b), the Secretary of Defense shall submit to Congress a report on new CTR initiatives. The report shall include—

(A) a summary of the results of the study carried out under subsection (b);

(B) an assessment by the Secretary of the study; and

(C) a statement of the actions, if any, to be undertaken by the Secretary to implement any recommendations in the study.

(2) FORM- The report shall be in unclassified form but may include a classified annex if necessary.

(d) Funding- Of the amounts appropriated pursuant to the authorization of appropriations in section 301(19) or otherwise made available for Cooperative Threat Reduction programs for fiscal year 2008, not more than $1,000,000 shall be obligated or expended to carry out this section.

Appendix B

Biographical Sketches of Committee Members

David R. Franz, *Co-chair*, served in the U.S. Army Medical Research and Materiel Command for 23 of 27 years on active duty and retired as colonel. He served as Commander of the U.S. Army Medical Research Institute of Infectious Diseases and as Deputy Commander of the Medical Research and Materiel Command. Prior to joining the command, he served as group veterinarian for the 10th Special Forces Group (Airborne). Dr. Franz was the chief inspector on three United Nations Special Commission biological warfare inspection missions to Iraq and served as technical advisor on long-term monitoring. He also served as a member of the first two U.S.-UK teams that visited Russia in support of the Trilateral Joint Statement on Biological Weapons and as a member of the Trilateral Experts' Committee for biological weapons negotiations. Dr. Franz was technical editor for the *Textbook of Military Medicine on Medical Aspects of Chemical and Biological Warfare* released in 1997. Current standing committee appointments include the Defense Intelligence Agency Red Team Bio-Chem 2020, the Defense Threat Reduction Agency's Threat Reduction Advisory Committee, the Department of Health and Human Services' National Science Advisory Board for Biosecurity, the Sandia National Laboratories' Distinguished Advisory Panel for international activities, and the Department of Homeland Security's Science and Technology Advisory Committee. He serves on the boards of the Federation of American Scientists and the Kansas Bioscience Authority. Dr. Franz holds an adjunct appointment as professor for the Department of Diagnostic Medicine and Pathobiology at the College of Veterinary Medicine, Kansas State University, and serves on the Dean's Advisory Council. The current focus of his activities relates to the role of international engagement in the life sciences as a component of national security policy. Dr. Franz holds a D.V.M. from Kansas State University and a Ph.D. in physiology from Baylor College of Medicine.

Ronald F. Lehman II, *Co-chair,* is the director of the Center for Global Security Research at Lawrence Livermore National Laboratory (LLNL) and also chairman of the Governing Board of the International Science and Technology Center. He serves on the Department of Defense (DOD) Threat Reduction Advisory Committee, and served on the Defense Science Board Task Forces on Globalization and Security and on Tritium, and on the National Research Council's Committee on Science, Technology, and Health Aspects of the Foreign Policy Agenda of the United States. In 1995, President William J. Clinton appointed him to the five-member President's Advisory Board on Arms Proliferation Policy. From 1989 to 1993, he was director of the U.S. Arms Control and Disarmament Agency. Previously, he served as assistant secretary of defense for International Security Policy, Department of State's U.S. chief negotiator on Strategic Offensive Arms, and deputy assistant to the President for National Security Affairs. He has also served as a senior director at the National Security Council (NSC), and senior professional staff of the Senate Armed Services Committee (SASC). Additionally, he headed the U.S. Delegations to the Fourth Review Conference of the Nonproliferation Treaty and the Third Review Conference of the Biological Weapons Convention, and also served as deputy head of delegation for the Chemical Weapons Convention signing.

Robert B. Barker retired from LLNL in 1999 after 30 years of service. He was a nuclear weapons designer and held several managerial positions, including assistant to the director. From 1986 to 1992, he served as assistant to the secretary of defense for Atomic Energy. Prior to this, he was deputy assistant director of the Bureau of Verification and Intelligence at the Arms Control and Disarmament Agency, 1983-1985. Dr. Barker also worked at the LLNL as assistant associate director for arms control, 1982-1983; special projects division leader, 1978-1982; and evaluation and planning division leader, 1973-1978. He has also served on the National Security Advisory Council.

William F. Burns (U.S. Army Major General, retired) is a former director of the U.S. Arms Control and Disarmament Agency and former commandant of the U.S. Army War College. He led the U.S. delegation on Safety, Security, and Dismantlement of nuclear weapons, serving as ambassador in negotiations on the denuclearization of the former Soviet Union. He is a distinguished fellow at the Army War College and serves on several panels, advisory boards, and boards of trustees of governmental and nonprofit organizations. He is judge emeritus of the Court of Judicial Discipline of Pennsylvania. General Burns co-chaired a National Academies' study on overcoming impediments to U.S.-Russian cooperation on nuclear nonproliferation and retired at the end of 2007 from the Committee on International Security and Arms Control (CISAC).

Rose Gottemoeller served as director of the Carnegie Moscow Center from January 2006 through December 2008. She was previously a senior associate at the Carnegie Endowment for International Peace, specializing in arms control, nonproliferation, and nuclear security issues. From 1998 to 2000, she served in the Department of Energy as assistant secretary for nonproliferation and national security and then as deputy undersecretary for defense nuclear nonproliferation. From 1993 to 1994, she was director for Russia, Ukraine, and Eurasia Affairs on the NSC in the White House. Ms. Gottemoeller co-chaired a National Academies' study on overcoming impediments to U.S.-Russian cooperation on nuclear nonproliferation and is currently a member of CISAC.

John Hamre was elected president and chief executive officer of the Center for Strategic and International Studies (CSIS) in January 2000. Before joining CSIS, he served as the 26th U.S. deputy secretary of defense. Prior to that, from 1993 to 1997, he served as under secretary of defense (comptroller). As comptroller, he was the principal assistant to the secretary of defense for the preparation, presentation, and execution of the defense budget and management improvement programs. In 2007, Secretary of Defense Robert Gates appointed Dr. Hamre to serve as chairman of the Defense Policy Board. Before serving at DOD, Dr. Hamre worked for 10 years as a professional staff member of the SASC. During that time, he was primarily responsible for the oversight and evaluation of procurement, research, and development programs, defense budget issues, and relations with the Senate Appropriations Committee. From 1978 to 1984, Dr. Hamre served in the Congressional Budget Office, where he became its deputy assistant director for national security and international affairs. In that position, he oversaw analysis and other support for committees in both the House of Representatives and the Senate.

Robert Joseph is currently a senior scholar at the National Institute for Public Policy. From June 2005 to March 2007, Ambassador Joseph served as the undersecretary of state for arms control and international security. Previously, he served as special assistant to the President and senior director for proliferation strategy, counterproliferation, and homeland defense NSC. From 1992 until 2001, Dr. Joseph was professor of national security studies and director-founder of the Center for Counterproliferation Research at the National Defense University. Before that, he was U.S. commissioner to the Standing Consultative Commission and to the U.S.-Russian Consultative Commission on Nuclear Testing, principal deputy assistant secretary of defense for international security policy, and deputy assistant secretary for nuclear forces and arms control policy. Dr. Joseph received his M.A. from the University of Chicago and his Ph.D. from Columbia University.

Orde Kittrie is a tenured professor of law at Arizona State University (ASU). Before joining the ASU law faculty in 2004, Professor Kittrie spent 11 years at the Department of State, including service as the State Department's senior attorney for nuclear affairs, as director of the office of international anticrime programs, as an attorney specializing in arms and dual-use trade controls, and as special assistant to the undersecretary for economic and business affairs. As senior attorney for nuclear affairs, he participated in negotiation of five nuclear nonproliferation agreements between the United States and Russia, served as counsel for the U.S. government's sanctions and other responses to the 1998 Indian and Pakistani nuclear tests, and helped negotiate at the United Nations Convention for the Suppression of Acts of Nuclear Terrorism. In 2005, Kittrie served as a member of the National Academy of Sciences' committee that produced with the Russian Academy of Sciences (RAS) a joint report entitled *Strengthening U.S.-Russian Cooperation on Nuclear Nonproliferation*. Kittrie currently serves as chair of the Nonproliferation, Arms Control and Disarmament Committee of the American Branch of the International Law Association; as chair of the Nonproliferation, Arms Control and Disarmament Committee of the American Society of International Law; as a visiting scholar at the Johns Hopkins University School of Advanced International Studies; and as a life member of the Council on Foreign Relations.

James LeDuc directs the Program on Global Health within the Institute for Human Infections and Immunity at the University of Texas Medical Branch. He also serves as deputy director of the Galveston National Laboratory. Previously he served as the Coordinator for Influenza for the Centers for Disease Control and Prevention (CDC) in Atlanta, Georgia, and was the director of the Division of Viral and Rickettsial Diseases in the National Center for Infectious Diseases (NCID), CDC. His professional career began as a field biologist working with the Smithsonian Institution's African Mammal Project in West Africa. Following that he served for 23 years as an Officer with the United States Army Medical Research and Development Command. He joined CDC in 1992, and was assigned to the World Health Organization as a Medical Officer, later becoming the Associate Director for Global Health at NCID. His research interests include the epidemiology of arboviruses and viral hemorrhagic fevers, and global health. He has participated in a number of National Research Council studies.

Richard W. Mies (U.S. Navy Admiral, retired) is currently a private consultant. He was previously the president and chief executive officer of Hicks and Associates, Inc., and was concurrently the deputy group manager of the Transformation, Training, Test, and Logistics Group at Science Applications International Corporation (SAIC). Admiral Mies joined SAIC after retiring from the U.S. Navy in February 2002 at the rank of admiral. During his military career, Admiral Mies served as commander in chief, United States Strategic Command, and in several senior staff positions. His many service decorations include the Defense Distinguished Service Medal, Navy Distinguished Service Medal, Defense Superior Service Medal (two awards), Legion of Merit (four awards), and National Intelligence Distinguished Service Medal. Admiral Mies graduated from the U.S. Naval Academy with a B.S. and holds a masters' degree in government administration and international relations and an honorary doctorate of law degree from the University of Nebraska.

Judith Miller is an author and a Pulitzer Prize-winning investigative reporter formerly with the *New York Times*. She left the paper in November 2005, after spending 85 days in jail to defend a reporter's right to protect confidential sources. In 2007, she joined the Manhattan Institute as an adjunct fellow and a contributing editor of the Manhattan Institute's *City Journal*. She writes for several publications–the *Wall Street Journal*, *Los Angeles Times*, and *New York Sun*, among them. She is also a commentator for Fox News on national security, focusing on the Middle East and counterterrorism, and the need to strike a delicate balance between protecting both national security and American civil liberties in a post-9/11 world. She has reported extensively on cooperative threat reduction activities, particularly in Russia. She is the author/coauthor of four books, and in 2002, was part of a small team that won a Pulitzer Prize for explanatory journalism for her January 2001 series on Osama bin Laden and al Qaeda. That same year, she won an Emmy for her work on a Nova–*New York Times* documentary based on articles for her book *Germs*. She was also part of the *Times* team that won the prestigious DuPont award for a series of programs on terrorism for Public Broadcasting Service's "Frontline." She has a B.A. from Barnard College and an M.P.A. from Princeton University's Woodrow Wilson School.

George W. Parshall is an advisor to the U.S. Army on neutralization processes used to destroy chemical weapons instead of incineration. Now retired, he joined DuPont's Central Research Department in 1954, where he rose to director of chemical science. He directed the work of 50 to 100 DuPont scientists, including that of Richard Schrock, who received the 2005 Nobel Prize in Chemistry. He was most closely associated with the DuPont processes for making critical intermediates used in producing nylon, polyester, and spandex polymers as well as alternatives to chlorofluorocarbon refrigerants. He coauthored the definitive textbook on "Homogeneous Catalysis." In the 1970s, he played a role in technological exchanges with the RAS. More recently, under the Department of Defense Cooperative Threat Reduction Program, he helped assess Russian technology for the destruction of chemical weapons. Parshall is a member of the American Academy of Arts and Sciences and the National Academy of Sciences.

Thomas Pickering (U.S. Ambassador, retired) is vice chairman of the consulting firm Hills & Company. He is the former senior vice president for international relations at the Boeing Company, a position he assumed in January 2001 upon his retirement as U.S. undersecretary of state for political affairs. Ambassador Pickering held the personal rank of career ambassador, the highest in the U.S. Foreign Service. In a diplomatic career spanning five decades, he has served as U.S. ambassador to the Russian Federation, India, Israel, El Salvador, Nigeria, and the Hashemite Kingdom of Jordan. From 1989 to 1993, he served as ambassador to the United Nations. His service in the U.S. government began in 1956 in the U.S. Navy. On active duty until 1959, he later served in the Naval Reserve to the grade of lieutenant commander. Between 1959 and 1961, he served in the Bureau of Intelligence and Research of the State Department, and in the Arms Control and Disarmament Agency. Ambassador Pickering previously served on the National Academies' Policy and Global Affairs Committee.

Kim Savit is currently a consultant for the Intelligence, Security and Technology Group of SAIC. She is also an adjunct professor at the University of Denver Korbel Graduate School of International Studies. Ms. Savit retired in May 2006 from her position as the senior professional staff member for the Middle East, Central and South Asia on the Majority Staff of the United States Senate Foreign Relations Committee. Ms. Savit served in the State Department as the deputy coordinator for security and law enforcement assistance to Europe and Eurasia (Acting, 2002-2003) and as the director for security and law enforcement assistance to the Newly Independent States of the former Soviet Union (1995-2002). Ms. Savit held many positions in DOD, including director of the Cooperative Threat Reduction Program for the Office of the Secretary of Defense and country desk officer for Morocco, Tunisia, Algeria, Libya, Iran, and Iraq in the Office of the Secretary of Defense, Near East and South Asian Affairs Bureau.

Appendix C

Department of Defense Cooperative Threat Reduction Program History: References

Allison, Graham, Ashton B. Carter, Steven E. Miller, and Philip Zelikow, eds. 1993. *Cooperative Denuclearization: From Pledges to Deeds*. Cambridge, Mass.: Center for Science and International Affairs, Harvard University, January 1993.

Allison, Graham, Owen R. Cote, Jr., Richard A. Falkenrath, and Steven E. Miller. 1996. *Avoiding Nuclear Anarchy: Containing the Threat of Loose Russian Nuclear Weapons and Fissile Material*. Cambridge, Mass.: Massachusetts Institute of Technology (MIT) Press. Accessed at http://mitpress.mit.edu/catalog/item/default.asp?ttype=2&tid=5530 on May 18, 2009.

Binnendijk, Hans and Mary Locke. 1993. *The Diplomatic Record, 1991-1992*. Boulder, CO: Westview Press.

Carnegie Endowment for International Peace. 2002. *Re-Shaping U.S.–Russian Threat Reduction: New Approaches for the Second Decade*. Washington, D.C.: Carnegie Endowment for International Peace and the Russian American Nuclear Security Advisory Council. Accessed at http://www.carnegieendowment.org/files/ReshapingThreatReduction.pdf on May 18, 2009.

Carter, Ashton and William J. Perry. 1999. *Preventive Defense: A New Security Strategy for America*. Washington, D.C.: Brookings Institution Press.

Carter, Ashton, William J. Perry, and John D. Steinbruner. 1992. *A New Concept of Cooperative Security*. Washington, D.C.: Brookings Institution Press.

Carter, Ashton, Kurt Campbell, Steven Miller, and Charles Zraket. 1991. *Soviet Nuclear Fission: Control of the Nuclear Arsenal in a Disintegrating Soviet Union*. Cambridge, Mass.: Center for Science and International Affairs, Harvard University.

Einhorn, Robert and Michèle Flournoy. *Protecting Against the Spread of Nuclear, Biological, and Chemical Weapons: An Action Agenda for the Global Partnership*. 2003. Washington, D.C.: Nuclear Threat Initiative.

Moltz, James Clay, et al. Special Report: Assessing U.S. Nonproliferation Assistance to the Newly Independent States. 2000. *The Nonproliferation Review*. 7:1. Accessed at http://cns.miis.edu/npr/71toc.htm on May 18, 2009.

Orlov, Vladimir, Roland Timerbaev, and Anton Kholpkov. 2002. *Nuclear Nonproliferation in U.S.-Russian Relations: Challenges and Opportunities*. Moscow, Russia: PIR Center Library Series. Accessed at http://www.isn.ethz.ch/isn/Digital-Library/Publications/Detail/?ots591=0C54E3B3-1E9C-BE1E-2C24-A6A8C7060233&lng=en&id=54762 on May 18, 2009.

Reiss, Mitchell. *Bridled Ambition: Why Countries Constrain Their Nuclear Capabilities*. 1995. Baltimore, Md.: Woodrow Wilson Center Press.

Shields, John. 1995. Reports: CIS Nonproliferation Developments. *The Nonproliferation Review.* 3:1. Accessed at http://cns.miis.edu/npr/31toc.htm on May 18, 2009.
Shields, John and William C. Potter, eds. 1997. *Dismantling the Cold War: U.S. and Newly-Independent States Perspectives on the Nunn-Lugar Cooperative Threat Reduction Program.* Cambridge, Mass.: MIT Press. Accessed at http://mitpress.mit.edu/catalog/item/default.asp?ttype=2&tid=3868 on May 18, 2009.

Appendix D

List of Committee Meetings and Speakers

Committee Meeting #1: May 21, 2008, Washington, D.C.

Speakers
Joseph Benkert, Department of Defense
Joseph P. Harahan, Defense Threat Reduction Agency (DTRA)
Mary Alice Hayward, Department of State
Kenneth Luongo, Partnership for Global Security
Mary Beth Dunham Nikitin, Congressional Research Service
Jason Rao, Department of State
Sharron Squassoni, Carnegie Endowment for International Peace
Amy Smithson, Monterey Institute of International Studies
James Tegnelia, DTRA
Charles Thornton, University of Maryland
William Tobey, Department of Energy
Elizabeth Turpen, The Henry L. Stimson Center

Committee Meeting #2: July 8, 2008, Washington, D.C.

Speakers
Joe DeThomas, Civilian Research and Development Corporation
Susan Koch, Department of State
Charles Lutes, National Security Council
Neile Miller, Office of Management and Budget
William Steiger, Health and Human Services

Committee Meeting #3: October 9, 2008, Washington, D.C.
Writing Meeting

Committee Meeting #4: November 17, 2008, Washington, D.C.
Writing meeting

Appendix E

The Evolution of U.S. Government Threat Reduction Programs

Beginning in 2002, the annual legislative process included language in authorization bills for the Departments of Defense (DOD), State, and Energy (DOE) that reflected the changing nature of the threat environment and shifting security priorities. In particular, Congress expressed an interest in seeing U.S. government threat reduction programs diversify beyond their previous geographic boundary of the former Soviet Union (FSU) and become more relevant to post-September 11, 2001, efforts to counter terrorism and prevent possible terrorist acquisition of weapons of mass destruction materials, technologies, and expertise. The legislative evolution is summarized below by fiscal year starting in 2002.

Fiscal Year	Agency	Congressional Action
2002	General	Administration failed to certify Russia for DOD Cooperative Threat Reduction (CTR) funding because it could not certify Russia's compliance with its obligations under the Biological and Chemical Weapons Conventions. This finding (which continued in subsequent years) delayed several ongoing programs and required a waiver before programs could proceed. The program disruption affected principally DOD's programs and a few State Department efforts; most DOE and State Department efforts were unaffected. Following negotiations with Congress, annual waiver authority was granted for the next 3 years.
	DOD	Congress asked that within 180 days of enactment of the National Defense Authorization Act that DOD produce "a report to Congress describing the steps that have been taken to develop cooperative threat reduction programs with India and Pakistan." Such a report was to include "recommendations for changes in any provision of existing law that was an impediment to the full establishment of such programs, a timetable for implementation of such programs, and an estimated 5-year budget that would be required to fully fund such programs" (HR 3338). There is no evidence that this report was ever produced.
2003	DOS	The omnibus appropriations bill authorized the State Department to use Nonproliferation and Disarmament Funds (NDF) "for such countries other than the Independent States of the FSU and international organizations when it is in the national security interest of the United States to do so" (H.J. Res. 2). This language was maintained in subsequent years.
		The State Department had already used NDF for transporting highly enriched uranium out of Belgrade, and by 2004 would be using it to aid in Libyan Weapons of Mass Destruction (WMD) dismantlement. Similarly, the State Department's program to redirect former weapon scientists began work in Iraq in 2003 and in Libya in 2004.

Fiscal Year	Agency	Congressional Action
2004	**DOD**	Congress explicitly authorized use of DOD CTR funds outside the FSU, provided the project "will assist the United States in the resolution of a critical emerging proliferation threat or permit the United States to take advantage of opportunities to achieve long-standing nonproliferation goals, can be completed in a short period of time, and that the DOD is the entity of the Federal Government that is most capable of carrying out such project or activity" (HR 1588). No specific suggestions were given.
	DOE	Congress authorized DOE to use international nuclear materials protection and cooperation program funds outside the former Soviet Union if it "will assist the United States in the resolution of a critical emerging proliferation threat or permit the United States to take advantage of opportunities to achieve long-standing nonproliferation goals, can be completed in a short period of time, and that the Department of Energy is the entity of the Federal Government that is most capable of carrying out such project or activity" (HR 1588).
		DOE soon organized its global nonproliferation efforts into the Global Threat Reduction Initiative and joined with the State Department in an effort to redirect former weapons scientists in Iraq and Libya.
2005	**DOE**	Congress further emphasized the global role of DOE's nonproliferation programs. The defense authorization bill stated the Sense of Congress that "the security, including the rapid removal or secure storage, of high-risk, proliferation-attractive fissile materials, radiological materials, and related equipment at vulnerable sites worldwide should be a top priority among the activities to achieve the national security of the United States." It then authorized a range of nonproliferation efforts, including scientist redirection, that are global in scope. In addition, it proposed a DOE pilot program in Georgia, called the Silk Road Initiative, to redirect weapons of mass destruction scientists. Eventually, it may include a much wider range of former Soviet republics: Armenia, Azerbaijan, Georgia, Kazakhstan, Kyrgyzstan, Tajikistan, Turkmenistan, and Uzbekistan.

Fiscal Year	Agency	Congressional Action
2006	General	Congress requested a report from the President by November 2006, that details "impediments to the effective conduct of CTR programs and related threat reduction and nonproliferation programs and activities in the states of the FSU" and steps necessary to overcome them (Public Law 109-163).
	DOD	The 2005 National Academies' report *Strengthening U.S.-Russian Cooperating on Nuclear Nonproliferation* finds that "[t]he U.S. government's ability to provide nonproliferation assistance to Russia has at times been severely complicated by legislative requirements stipulating that the President must certify that Russia has met standards that, in some cases, have little connection to the assistance in question. . . . The joint committee recommends that the U.S. Congress either repeal such certification requirements or provide the President with permanent waiver authority." The defense authorization bill included permanent waiver authority, but the President must still present a waiver each year if he cannot certify Russia's compliance with the requirements, but this authority is available to him every year. Senator Richard Lugar attempted to broaden the bill to further encourage work outside the FSU, but those provisions were not included in the final version.
	DOS	The State Department launched its Global Biosecurity Program aimed at countries in Middle East, South and East Asia, and the Pacific. In fiscal year (FY) 2006, projects began in Indonesia, Pakistan, and the Philippines. By FY 2007, it was working in Egypt and Yemen, with contacts also in Latin America. The program focuses on laboratory security, pathogen consolidation and security, and biosafety.
2007	DOD	Waiver authority extended.

Fiscal Year	Agency	Congressional Action
2008	**General**	The 9/11 Commission Act recommended the following: "The United States should expand, improve, increase resources for, and otherwise fully support the CTR program." The Sense of Congress was that future funding should be increased and programs accelerated (HR 1).
		The act also authorized increased and accelerated funds for DOE various nonproliferation efforts.
	DOD	Congress finally repealed the Presidential certification requirements.
		The act provides $10 million explicitly for new CTR initiatives that are outside the FSU (See Annex xxx, Title XIII of act), included a list of principles that should guide new initiatives, and suggested that new initiatives be considered in Asia, the Middle East, and that related to the denuclearization of the Democratic People's Republic of Korea.
		The conference report directed $5 million of the funds appropriated under chemical weapons destruction be made available as initial funding for a chemical weapons incinerator in Libya, pending authorization for such activity.

Appendix F

Nunn-Lugar Scorecard[1]

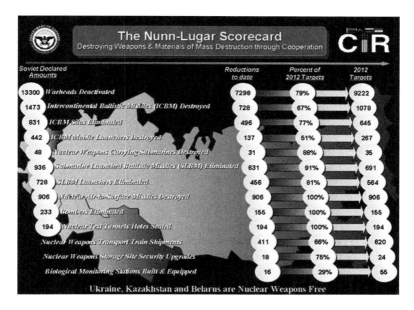

Nunn-Lugar definition of terms:
- ICBM – Intercontinental ballistic missile
- SLBM – Submarine-launched ballistic missile
- SSBN – Nuclear submarine capable of launching ballistic missile
- ASM – Air-to-surface missile

---

[1] As of December 17, 2008. Available at http://lugar.senate.gov/nunnlugar/scorecard.html.

Appendix G

The G8 Global Partnership: Guidelines for New or Expanded Cooperation Projects[1]

The G8 will work in partnership, bilaterally and multilaterally, to develop, coordinate, implement and finance, according to their respective means, new or expanded cooperation projects to address (i) nonproliferation, (ii) disarmament, (iii) counter-terrorism and (iv) nuclear safety (including environmental) issues, with a view to enhancing strategic stability, consonant with our international security objectives and in support of the multilateral nonproliferation regimes. Each country has primary responsibility for implementing its nonproliferation, disarmament, counter-terrorism and nuclear safety obligations and requirements and commits its full cooperation within the Partnership.

Cooperation projects under this initiative will be decided and implemented, taking into account international obligations and domestic laws of participating partners, within appropriate bilateral and multilateral legal frameworks that should, as necessary, include the following elements:

i. Mutually agreed effective monitoring, auditing, and transparency measures and procedures will be required in order to ensure that cooperative activities meet agreed objectives (including irreversibility as necessary), to confirm work performance, to account for the funds expended and to provide for adequate access for donor representatives to work sites;

ii. The projects will be implemented in an environmentally sound manner and will maintain the highest appropriate level of safety;

---

[1] Statement by G8 Leaders. The G8 Global Partnership Against the Spread of Weapons and Materials of Mass Destruction. Kananasis Summit. 2002. Available as of March 2009 at htttp://www.dt.tesoro.it/opencms/opencms/handle404?exporturi=/export/sites/sitodt/modules/documenti_it/rapporti_finanziari_internazionali/rapporti_finanziari_internazionali/G8_on_Global_partnership.pdf&%5d

iii. Clearly defined milestones will be developed for each project, including the option of suspending or terminating a project if the milestones are not met;

iv. The material, equipment, technology, services and expertise provided will be solely for peaceful purposes and, unless otherwise agreed, will be used only for the purposes of implementing the projects and will not be transferred. Adequate measures of physical protection will also be applied to prevent theft or sabotage;

v. All governments will take necessary steps to ensure that the support provided will be considered free technical assistance and will be exempt from taxes, duties, levies, and other charges;

vi. Procurement of goods and services will be conducted in accordance with open international practices to the extent possible, consistent with national security requirements;

vii. All governments will take necessary steps to ensure that adequate liability protections from claims related to the cooperation will be provided for donor countries and their personnel and contractors;

viii. Appropriate privileges and immunities will be provided for government donor representatives working on cooperation projects; and

ix. Measures will be put in place to ensure effective protection of sensitive information and intellectual property.

Given the breadth and scope of the activities to be undertaken, the G8 will establish an appropriate mechanism for the annual review of progress under this initiative which may include consultations regarding priorities, identification of project gaps and potential overlap, and assessment of consistency of the cooperation projects with international security obligations and objectives. Specific bilateral and multilateral project implementation will be coordinated subject to arrangements appropriate to that project, including existing mechanisms.

For the purposes of these guidelines, the phrase "new or expanded cooperation projects" is defined as cooperation projects that will be initiated or enhanced on the basis of this Global Partnership. All funds disbursed or released after its announcement would be included in the total of committed resources. A range of financing options, including the option of bilateral debt for program exchanges, will be available to countries that contribute to this Global Partnership.

The Global Partnership's initial geographic focus will be on projects in Russia, which maintains primary responsibility for implementing its obligations and requirements within the Partnership.

In addition, the G8 would be willing to enter into negotiations with any other recipient countries, including those of the former Soviet Union, prepared to adopt the guidelines, for inclusion in the Partnership.

Recognizing that the Global Partnership is designed to enhance international security and safety, the G8 invites others to contribute to and join in this initiative.

With respect to nuclear safety and security, the partners agreed to establish a new G8 Nuclear Safety and Security Group by the time of our next Summit.

Appendix H

A Comparison of the Characteristics of Six Weapons Systems from the Perspective of a State or Terrorist Organization[1]

Legend

Detectable signature?	No, Maybe, Yes
Barriers to development	Very High, High, Moderate, Low
Technical experts	Many, Some, Few, Very Few
Casualty potential	Very High, High, Moderate, Low, Low (High Economic), Very Low
Is technical capability a limitation?	Yes, No
Availability of materials	Few, Some, Many
Access to materials	Very Poor, Poor, Good, Very Good, Universal
Intelligence target	Very Hard, Hard, Moderately Hard, Moderately Easy
Dual use	Yes, No
Trackable/detectable?	Yes, Maybe
Attribution possibility?	High, Moderate, Low, Very low

---

[1] Table developed by Dr. David Franz.

	Nuclear		Biological		Chemical	
	State	*Substate*	*State*	*Substate*	*State*	*Substate*
Detectable signature?	Yes	Maybe	Maybe	Maybe	Yes	Maybe
Barriers to development	High	Very High	Moderate	Moderate	Low	Moderate
Technical experts	Few	Very Few	Many	Very Few	Many	Few
Casualty potential	Very High	Very High	Very High to Very Low	High to Very Low	Moderate to Low	Low
Is technical capability a limitation?	Yes	Yes	No	Yes	No	No
Availability of materials	Few	Few	Many	Many	Some	Some
Access to materials	Poor	Very Poor	Good	Poor	Very Good	Good
Intelligence target	Moderately Easy	Moderately Hard	Very Hard	Very Hard	Moderately Hard	Hard
Dual use	Yes	N/A	Yes	N/A	Yes	N/A
Trackable/detectable?	Yes	Maybe	Maybe	Maybe	Maybe	Maybe
Attribution possibility?	High to Moderate	Moderate	Low	Very Low	Low	Very Low

	Radiological		Conventional Explosive		Cyber	
	State	*Substate*	*State*	*Substate*	*State*	*Substate*
Detectable signature?	Maybe	Maybe	No	No	Maybe	Maybe
Barriers to development	Low	Moderate	Low	Low	Low	Low
Technical experts	Some	Few	Many	Many	Many	Some
Casualty potential	Low (High Economic)	Low (High Economic)	High to Low	High to Low	N/A	N/A
Is technical capability a limitation?	No	Yes	No	No	No	No
Availability of materials	Some	Some	Many	Many	Universal	Universal
Access to materials	Good	Poor	Very Good	Good	Universal	Universal
Intelligence target	Very Hard	Hard	Very Hard	Very Hard	Hard	Hard
Dual Use	Yes	N/A	Yes	N/A	No	N/A
Trackable/detectable?	Maybe	Maybe	Maybe	Maybe	Maybe	Maybe
Attribution possibility?	Low	Very Low	Low	Very Low	Moderate	Moderate

Appendix I

Department of Defense Cooperative Threat Reduction Programs

PROGRAMS

The Cooperative Threat Reduction (CTR) program pursues four objectives to reduce the present threat of weapons of mass destruction (WMD) and guarantee national security.

Objective 1: Dismantle former Soviet Union (FSU) WMD and associated infrastructure
Objective 2: Consolidate and secure FSU WMD and related technology and materials
Objective 3: Increase transparency and encourage higher standards of conduct
Objective 4: Support defense and military cooperation with the objective of preventing proliferation

There are several programs aimed to meet each objective.

OBJECTIVE 1: DISMANTLE FSU WMD AND ASSOCIATED INFRASTRUCTURE

Strategic Offensive Arms Elimination (SOAE) Program–Russia: Department of Defense (DOD) continues to assist Russia by contracting for and overseeing destruction of strategic weapons delivery systems in accordance with the SOAE Implementing Agreement and applicable Strategic Arms Reduction Treaty (START) provisions, including the START Conversion or Elimination Protocol. CTR program assistance remains an incentive for Russia to draw down its Soviet-legacy nuclear forces, thereby reducing opportunities for their proliferation or use. DOD provides equipment and services to destroy or dismantle intercontinental ballistic missiles (ICBMs), ICBM silo launchers, road and rail mobile launchers, submarine-launched ballistic missiles (SLBMs), SLBM launchers, reactor cores of associated strategic nuclear-powered ballistic missile submarines, and WMD infrastructure. DOD also supports placement of spent fuel from naval nuclear reactors, referred to as Spent Naval Fuel, prior to its elimination, into casks designed for long-term storage as well as logistical and maintenance support for equipment.

Chemical Weapons Destruction (CWD) Program–Russia: In accordance with the CWD Implementing Agreement, DOD is assisting Russia with the safe, secure, and environmentally sound destruction of the most proliferable portion of its chemical weapons nerve-agent stockpile. The Chemical Weapons Destruction Facility and the former Chemical Weapons Production Facility demilitarization projects support this effort.

Strategic Nuclear Arms Elimination (SNAE) Program–Ukraine: CTR program assistance, consistent with the SNAE Implementing Agreement, includes elimination of Tu 22M Backfire and Tu-142 Bear nuclear-capable maritime patrol aircraft that are modifications of START-accountable heavy bombers, Kh 22 nuclear air-to-surface missiles, and strategic bomber trainers.

Weapons of Mass Destruction Infrastructure Elimination (WMDIE) Program–Ukraine: In accordance with the WMDIE Implementing Agreement, the Nuclear Weapons Storage Area project will eliminate infrastructure at sites formerly associated with nuclear weapons and warhead storage, operations, and maintenance that supported the forward-deployed nuclear weapons arsenals of the Soviet armed forces and assist in preventing the proliferation of associated design data, materials, equipment, and technologies.

Biological Threat Reduction Prevention (BTRP) Program–FSU: The BTRP program's objectives are to reduce the risk of bioterrorism and prevent the proliferation of biological weapons technology, expertise, and extremely dangerous pathogens (EDPs). The United States has CTR implementing agreements with Azerbaijan, Georgia, Kazakhstan, Ukraine, and Uzbekistan, to assist them in preventing the proliferation of biological weapons materials and expertise to rogue states and terrorist groups, increase transparency, encourage high standards of conduct by scientists, and preempt a "brain drain" of bio-related expertise. All BTRP projects in Russia fall under the International Science and Technology Center (ISTC) Agreement and the ISTC Funding Memorandum of Agreement. The U.S.–Kazakhstan WMDIE Implementing Agreement covers BTRP projects in Kazakhstan. Biological Threat Reduction Implementing Agreements have been signed with Azerbaijan, Georgia, Ukraine, and Uzbekistan. This program is executed through three projects, each of which serves a different objective of the CTR program:

1. **Biological Weapons Infrastructure Elimination** – Objective 1
2. **Biosecurity and Biosafety (BS&S) and Threat Agent Detection and Response (TADR) Network** – Objective 2
3. **Cooperative Biological Research** – Objective 3

OBJECTIVE 2: CONSOLIDATE AND SECURE FSU WMD AND RELATED TECHNOLOGY AND MATERIALS

Nuclear Weapons Storage Security (NWSS) Program–Russia: In accordance with the NWSS Implementing Agreement, this program helps support proliferation prevention by providing enhancements to the security systems of nuclear weapons storage sites.

The Personnel Reliability Program project was completed in August 2005, with delivery of the final 5,000 test cups. The Russian Ministry of Defense's 12th Main Directorate assumed full responsibility for the project.

Nuclear Weapons Transportation Security (NWTS) Program–Russia: In accordance with the NWTS Implementing Agreement, this program supports proliferation prevention by enhancing the security and safety of nuclear weapons during shipment. Much of the DOD-provided equipment is located at sensitive Ministry of Defense locations. It is shipped to less sensitive locations when DOD conducts audits and examinations.

Fissile Material Storage Facility (FMSF) Program–Russia: In accordance with the Fissile Material Storage Facility (FMSF) Construction Implementing Agreement, the facility will provide centralized, safe, secure, and ecologically sound storage for weapons-grade fissile material. The facility was completed and commissioned on December 11, 2003.

Biological Threat Reduction Prevention (BTRP) Program–FSU: DOD combined the BS&S and TADR programs into one project because of their close relationship and common objective. Their goals are to prevent the theft, sale, diversion, and accidental or intentional release of pathogens; consolidate pathogen collections and work at safe, secure centralized repositories; and strengthen the recipient states' detection and response networks for dangerous pathogens. Combining them enables a more integrated and streamlined approach to engaging institutes in the BTRP Program. BS&S-TADR efforts target dangerous pathogens that pose particular risks for theft, diversion, accidental release, or use by terrorists. In Russia, work is focused on BS&S enhancements, with no plans to create a TADR system.

OBJECTIVE 3: INCREASE TRANSPARENCY AND ENCOURAGE HIGHER STANDARDS OF CONDUCT

Biological Threat Reduction Prevention (BTRP) Program–FSU: Through the Cooperative Biological Research (CBR) program, DOD works with institutes and scientists previously involved in biological weapons research to employ them in peaceful research focusing on investigating dangerous pathogens for prophylactic, preventive, or other peaceful purposes. By so engaging former biological weapons scientists, CBR helps to prevent the proliferation of biological weapons scientific expertise and preempt potential "brain drain" of scientists to rogue states; increase the transparency at biological institutes and encourage higher standards of openness, ethics, and conduct by scientists; provide the United States access to expertise that can enhance preparedness against biological threats; enable the transfer of EDPs to the United States for study to improve public health; and enable forensics reference research.

OBJECTIVE 4: SUPPORT DEFENSE AND MILITARY COOPERATION WITH THE OBJECTIVE OF PREVENTING PROLIFERATION

Weapons of Mass Destruction Proliferation Prevention Initiative (WMD-PPI) Program–FSU, Except Russia: The WMD-PPI Program addresses the potential vulnerability of the non-Russian FSU states' borders to smuggling of WMD and related components. WMD-PPI attempts to complement the CTR program's traditional focus, WMD at its source, by addressing WMD on the move. Currently, DOD is helping Azerbaijan, Kazakhstan, Ukraine, and Uzbekistan to develop and sustain capabilities to prevent the proliferation of WMD-related materials, components, and technologies across their borders. Agreements are made with the recipient states to have them report any WMD detections made with U.S. government-supplied equipment to the in-country U.S. embassy, for forwarding to the U.S. government.

Defense and Military Contacts (DMC): The DMC program was created in 1993 as a part of the larger CTR program and attempts to develop positive relationships between the defense, military, and security communities of the United States and FSU states. Bilateral activities are designed to engage the military and defense officials of FSU states in activities that promote demilitarization and defense reform, further proliferation prevention efforts, and endorse regional stability and cooperation. The program is developed by the Office of the Assistant Secretary of Defense for International Security Policy, through the Office of the Deputy Assistant Secretary of Defense for Eurasia Policy, in close coordination with the Joint Staff, the Unified Combatant Commands, and the U.S. military services to ensure that scheduled events support the secretary of defense's Security Cooperation Guidance and regional commands' country and regional campaign plans.

Appendix J

Congressional Guidelines and Corresponding Findings and Recommendations

Congressional Guidelines	Corresponding Findings and Recommendations
1. Programs should be well coordinated with the Department of Energy, the Department of State, and any other relevant United States Government agency or department.	Findings 1-2; 2-5; 2-10; 3-2; 3-4; 4-2 Recommendations 1-1; 2-1; 3-1; 3-1a; 4-1
2. Programs will include appropriate transparency and accountability mechanisms, and legal frameworks and agreements between the United States and Cooperative Threat Reduction (CTR) partner countries.	Findings 1-5; 2-6; 3-1; 3-3; 3-5; 3-7 Recommendations 3-1b; 3-3; 3-3a; 3-3b; 3-3c
3. Programs should reflect engagement with nongovernment experts on possible new options for the CTR program.	Findings 2-7; 2-8; 2-9; 2-10; 2-11; 3-3; 3-4; 3-7 Recommendations 3-1a; 3-2; 4-1

163

4. Programs should include work with the Russian Federation and other countries to establish strong CTR partnerships. Among other things, these partnerships should:

(i) increase the role of scientists and government officials of CTR partner countries in designing CTR programs and projects;

(ii) increase financial contributions and additional commitments to CTR programs and projects from Russia and other partner countries, as appropriate, as evidence that the programs and projects reflect national priorities and will be sustainable.

Findings 1-2; 1-4; 2-4; 2-5; 2-6; 2-12; 3-4; 3-6; 3-8

Recommendations 3-1b; 3-2; 3-3b

5. Programs should include broader international cooperation and partnerships, and increased international contributions.

Findings 1-2; 1-4; 1-6; 2-2; 2-3; 2-4; 2-8; 2-11; 2-12; 3-3; 3-4; 3-6; 3-7

Recommendations 1-1; 3-1b

6. Programs should incorporate a strong focus on national programs and sustainability, which includes actions to address concerns raised and recommendations made by the Government Accountability Office, in its report of February 2007, titled "Progress Made in Improving Security at Russian Nuclear Sites, but the Long-Term Sustainability of U.S. Funded Security Upgrades is Uncertain," which pertain to the Department of Defense.

Findings 1-4; 2-3; 2-4; 2-12; 3-3; 3-4; 4-3

Recommendations 2-1; 3-3; 3-3c

7. Efforts should continue to focus on the development of CTR programs and projects that secure nuclear weapons; secure and eliminate chemical and biological weapons and weapons-related materials; and eliminate nuclear, chemical, and biological weapons-related delivery vehicles and infrastructure at the source.

Chapter 4

8. There should be efforts to develop new CTR programs and projects in Russia and the former Soviet Union, and in countries and regions outside the former Soviet Union, as appropriate and in the interest of United States national security.

Findings 1-1; 1-2; 1-3; 1-4; 1-6; 2-2; 2-3; 2-4; 2-8; 2-11; 2-12; 3-3; 3-4; 3-6; 3-7; 4-3

Recommendations 1-1; 3-1b